Ready, Set, Go!

A Student Guide to SPSS® 13.0 and 14.0 for Windows®

Thomas W. Pavkov

Kent A. Pierce

Purdue University Calumet

Boston   Burr Ridge, IL   Dubuque, IA   Madison, WI   New York
San Francisco   St. Louis   Bangkok   Bogotá   Caracas   Kuala Lumpur
Lisbon   London   Madrid   Mexico City   Milan   Montreal   New Delhi
Santiago   Seoul   Singapore   Sydney   Taipei   Toronto

Higher Education

READY, SET, GO! A STUDENT GUIDE TO SPSS 13.0 AND 14.0 FOR WINDOWS

Published by McGraw-Hill, a business unit of The McGraw-Hill Companies, Inc., 1221 Avenue of the Americas, New York, NY, 10020. Copyright 2007, 2003 by The McGraw-Hill Companies, Inc. All rights reserved. No part of this publication may be reproduced or distributed in any form or by any means, or stored in a database or retrieval system, without the prior written consent of The McGraw-Hill Companies, Inc., including, but not limited to, in any network or other electronic storage or transmission, or broadcast for distance learning.

Some ancillaries, including electronic and print components, may not be available to customers outside the United States.

This book is printed on acid-free paper.

1 2 3 4 5 6 7 8 9 0 DOC/DOC 0 9 8 7 6

ISBN-13: 978-0-07-312665-4
ISBN-10:   0-07-312665-9

Vice President and Editor-in-Chief: *Emily Barrosse*
Publisher: *Beth Mejia*
Executive Editor: *Michael J. Sugarman*
Editorial Coordinator: *Katherine C. Russillo*
Marketing Manager: *Melissa S. Caughlin*
Managing Editor: *Jean Dal Porto*
Project Manager: *Meghan Durko*
Art Director: *Jeanne Schreiber*
Designer: *Marianna Kinigakis*
Senior Media Producer: *Stephanie George*
Senior Supplement Producer: *Louis Swaim*
Composition: *11/14 Times, by GTS-India*
Printing: *50# Windsor Offset Smooth, R.R. Donnelley & Sons*

Library of Congress Control Number: 2006920655

The Internet addresses listed in the text were accurate at the time of publication. The inclusion of a Web site does not indicate an endorsement by the authors or McGraw-Hill, and McGraw-Hill does not guarantee the accuracy of the information presented at these sites.

www.mhhe.com

Preface

This handbook provides the basic information students need to use SPSS® for Windows® in both introductory statistics and research design courses. When used in conjunction with a primary statistics or research design textbook, this book is a flexible and up-to-date tool instructors can use to incorporate computerized statistical analysis into their courses.

This book emerged from our need to provide students with basic information on using SPSS for Windows, the statistical package that we use as part of our course of instruction in behavioral statistics. After searching without success for a guide that would efficiently and clearly provide this type of information, we developed our own instructional material for the course. *Ready, Set, Go! A Student Guide to SPSS® 6.1 for Windows®* was the result of our efforts. This handbook is an updated version designed for use with SPSS for Windows—Version 13.0 and 14.0 or later.

This handbook is designed to be an inexpensive source of "how-to" information for student users of the SPSS for Windows software. Each assignment provides the user with background information linking statistical methods and the SPSS procedures associated with those methods. The steps of these procedures are illustrated by numerous screen shots of the SPSS graphical user interface. The book also provides basic information on the interpretation of output produced by SPSS. Each chapter ends with an "On Your Own" section containing a learning task that encourages students to undertake independent computer assignments.

This book may be used for more than reference; each assignment has been developed as a guided exercise. Students are introduced to the research process as they work through exercises—formulating research questions, choosing appropriate statistical procedures, summarizing results, and interpreting data. We believe this approach will help students understand the research process and increase their confidence in using SPSS datasets of their choice.

The book is organized topically, covering most of the basic concepts presented in introductory statistics courses. We begin in Assignment 1 by providing the student user with information on how to access basic SPSS procedures such as loading data and printing output. Assignments 2 and 3 cover SPSS procedures for descriptive statistics and for the graphical presentation of data. Assignments 4 through 7 focus on using SPSS to compare groups and paired-samples *t* tests and then move to one-way analysis of variance for independent and related samples. Correlation and regression analysis are covered in Assignments 8 and 9. Finally, in Assignment 10, the student user is introduced to using SPSS for producing contingency tables and calculating the chi-square statistic. The Appendix contains information about alternate methods of data entry and some sample datasets.

We would like to acknowledge the colleagues, friends, and family members who have supported this project. In addition, we wish to thank Frank Graham for sponsoring this work in its infancy and our new editor Mike Sugarman. Thanks also to Meghan Durko for coordinating the final edit and production of this edition, and to Katherine Russillo, Editorial Coordinator; Marianna Kinigakis, Designer; Louis Swaim, Senior Production Supervisor; Melissa Caughlin, Marketing Manager; and Stephanie George, Senior Media Producer.

Contents

ASSIGNMENT 1

Learning the Basics of SPSS

OBJECTIVES

1. Load SPSS for Windows

2. Load a datafile into SPSS for Windows

3. Define SPSS variables

4. Obtain printouts from SPSS for Windows

5. Exit from SPSS for Windows

The purpose of this book is to provide general information for beginning users of SPSS for Windows–Version 13.0, 14.0, or later. This section of the book will provide you with skills you will need in order to use some of the basic procedures of SPSS for Windows. Access to SPSS for Windows varies from one computer facility to another and from computer to computer. As you use SPSS, you may encounter some issues not covered by this book. In that case, you should seek assistance from your instructor or statistical consultant. For the novice user of SPSS for Windows, some general issues need to be addressed. Of primary importance is starting the software application. Next, you will need to know how to load data from an existing datafile, define data, and perhaps enter data manually. You will also need to know how to print the results of your analysis following the completion of an SPSS procedure. These are the procedures you will learn in this section of the book.

STARTING SPSS

To start SPSS for Windows, you will need to start from your Windows Desktop (see Figure 1.1 for an example). Depending on your computer installation, you can access SPSS for Windows in different ways. You might start SPSS by double-clicking on a shortcut icon for SPSS on the Windows Desktop or by pointing to the SPSS icon in the program listings in the Windows Desktop Start menu (i.e., Start>Programs>SPSS for Windows>SPSS 14.0 for Windows). Please note that menu selection and ordering is identical for versions 13.0 and 14.0.

After you double-click on the SPSS icon, the computer will load the SPSS software. You will know SPSS is loading when the Windows hourglass replaces the

1

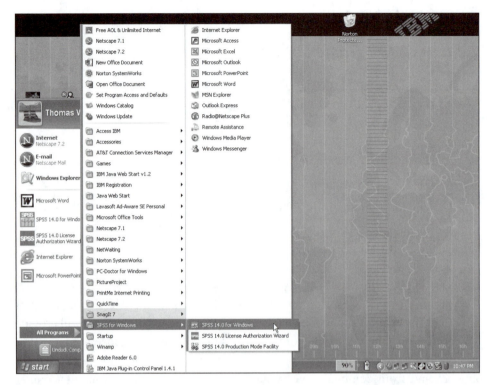

FIGURE 1.1   Using the Start Menu to Access SPSS 14.0 for Windows. Windows® XP is a registered trademark of Microsoft Corporation. Screen shot reprinted by permission from Microsoft Corporation.

pointer on your screen. The time required to load SPSS varies depending on the characteristics of your computer. These factors include the power of your machine, the location or type of installation, and the load on network resources. You can enhance the performance of the machine you are using by making sure that other Windows applications are not running simultaneously with SPSS for Windows. If you find other applications loaded, unloading them prior to running SPSS may enhance performance.

LOADING A FILE

After you initiate the loading of the SPSS software, one of two series of screens appears, allowing you to initiate the data loading process. One method of loading data involves using the File pull-down menu, as in previous versions of SPSS for Windows. A second method involves using an optional dialog query window. The method you use will depend on the way SPSS is installed at your location. Both methods are described in the following pages. If the dialog window is not in use at

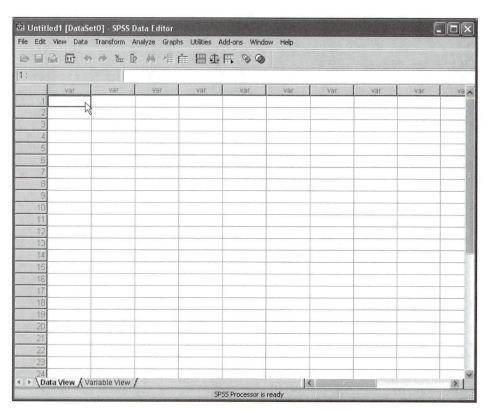

FIGURE 1.2   SPSS Data Editor Window

your location, a screen will appear that is similar to the screen shown in Figure 1.2. This screen is the main screen for SPSS for Windows and is called the SPSS Data Editor window. To load your data, point to the File menu, and highlight one of the file loading options appearing on the window.

Figure 1.3 displays the selections available at the beginning of an SPSS session from the File pull-down menu. You have the choice of creating a new file, opening an existing file, or reading data from an ASCII (text) format file. You can open the SPSS Syntax files to analyze problems with SPSS syntax. SPSS Syntax files are usually saved with the file extension .sps. These files contain the SPSS program language needed to run your SPSS assignments. You can also open SPSS Viewer files to examine and print the output from your SPSS assignments. These files are usually saved with the file extension .spo. SPSS Viewer files contain the information produced by SPSS statistical procedures. As shown in Figure 1.3, the most recently accessed SPSS datasets are listed as well. You can access one of these files by pointing and single-clicking on the highlighted name of the file. You will likely use the Open

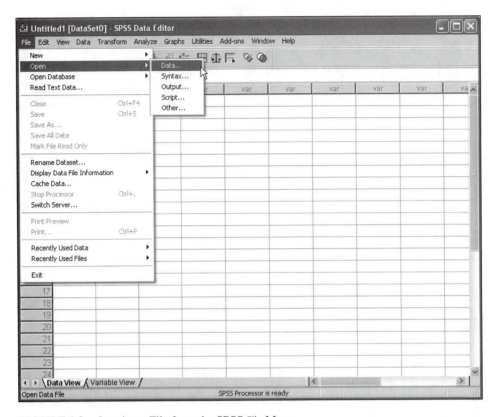

FIGURE 1.3   Opening a File from the SPSS File Menu

option most often. In the File pull-down menu, click on Open to open an existing file. In Figure 1.3, the Open choice is highlighted. If you choose the File>Open>Data option, the SPSS Open File window will appear next (see Figure 1.5).

The second method of opening a datafile involves using a Windows dialog box, as shown in Figure 1.4. If your installation uses the dialog option, this box will appear along with the SPSS Data Editor window in the background. This dialog window allows you to access the SPSS tutorial, type in data using the SPSS Data Editor, create or run a query using a database table, or open an existing SPSS datafile. If you choose to use this dialog box in the future, you are most likely to use the default Open an existing file option. Recently accessed files are listed in the small box under this option. You can access one of these files by highlighting and double-clicking on the file. SPSS will then load the file, and data will appear in the SPSS Data Editor window (see Figure 1.6). If the datafile you wish to access does not appear on the list, highlight and double-click on More Files. Then you can access your datafile using the Open File window, as illustrated in Figure 1.5. If you

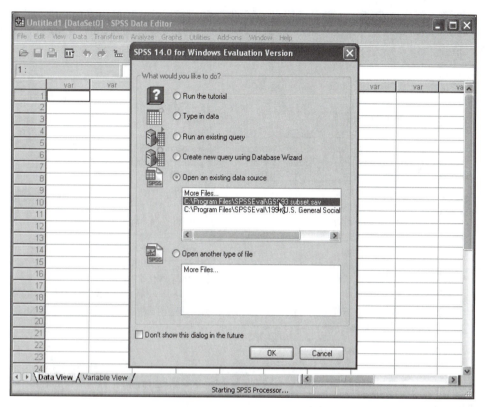

FIGURE 1.4 Opening a File from the SPSS for Windows Optional Dialog Box

wish to open a file that is not a data source, you may select the Open another type of file option. If you do not wish to use this dialog box in the future, click on the Don't show this dialog in the future box in the lower part of the dialog window.

USING THE Open File WINDOW

Once you select Open, the Open File window appears. Because thousands of files may be stored in the computer system, you will need to identify the location and name of the dataset you are going to use in the assignment. The Open File window allows you to direct SPSS in accessing named datasets. As shown in Figure 1.5, the Look in box at the top of the window shows the active folder. The active folder corresponds with the directory or subdirectory to which SPSS for Windows is pointed. You can change the directory by clicking on the down arrow at the right-hand side of the Look in box, clicking on the device or directory you want to access. Moving back through previously selected directories is accomplished by clicking on the Up One Level button to the right of the Look in box.

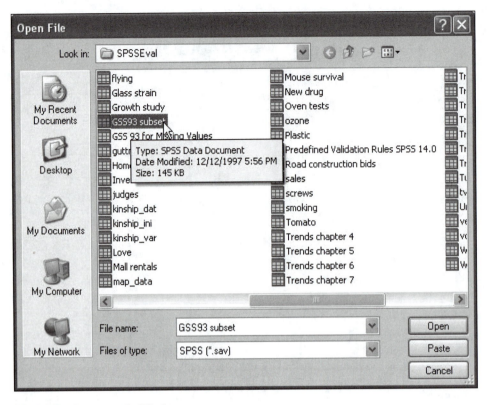

FIGURE 1.5   Open File Window

The large box in the middle of the Open File window contains the names of the files in the active folder. Figure 1.5 shows a list of files stored in the SPSS folder. To load one of these files, you must first highlight it and then point and click on the Open button or point and double-click on the highlighted file. Notice in Figure 1.5 that SPSS automatically looks for files with the extension .sav as designated in the Files of type box at the bottom of the Open File window. The .sav files are often referred to as SPSS "save" files and contain data stored in a compressed binary format. To complete the assignment, you will need to identify the appropriate .sav file, select that file, and load it. Your instructor will inform you which file (along with directory/folder location) to use as he or she makes each computer assignment. Once you locate the correct folder (see Figure 1.5), SPSS will automatically list the .sav files in the large box in the middle of the window.

In this example, the datafile named GSS93 subset is highlighted. This datafile will be used for illustration in many of the assignments throughout this book. The data in the GSS93 subset comes from the General Social Survey (GSS), administered periodically (recently every two years) by the National Opinion

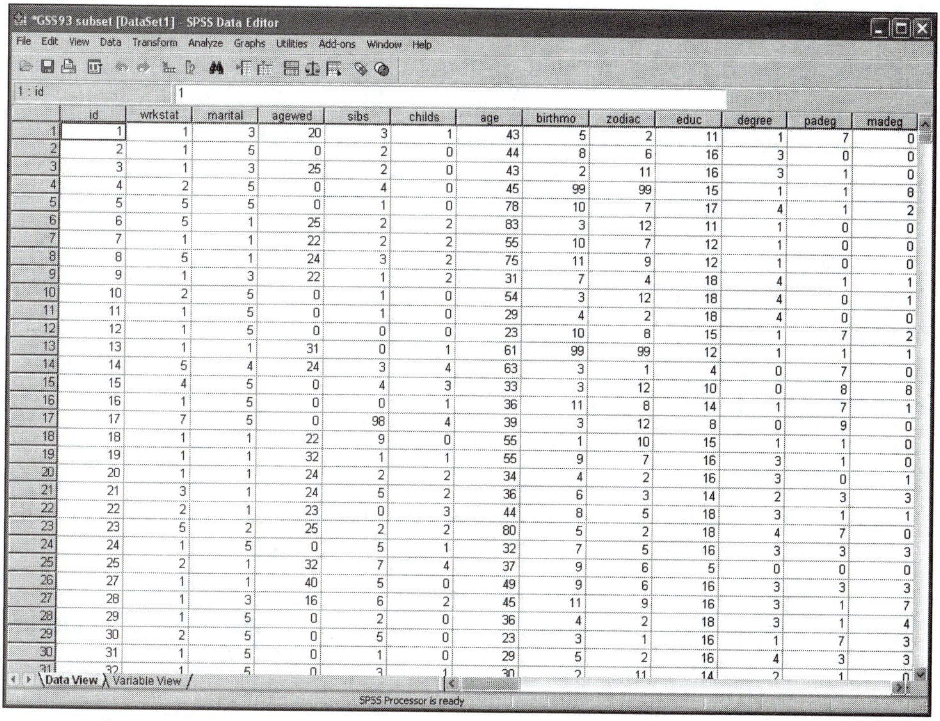

	id	wrkstat	marital	agewed	sibs	childs	age	birthmo	zodiac	educ	degree	padeg	madeg
1	1	1	3	20	3	1	43	5	2	11	1	7	0
2	2	1	5	0	2	0	44	8	6	16	3	0	0
3	3	1	3	25	2	0	43	2	11	16	3	1	0
4	4	2	5	0	4	0	45	99	99	15	1	1	8
5	5	5	5	0	1	0	78	10	7	17	4	1	2
6	6	5	1	25	2	2	83	3	12	11	1	0	0
7	7	1	1	22	2	2	55	10	7	12	1	0	0
8	8	5	1	24	3	2	75	11	9	12	1	0	0
9	9	1	3	22	1	2	31	7	4	18	4	1	1
10	10	2	5	0	1	0	54	3	12	18	4	0	1
11	11	1	5	0	1	0	29	4	2	18	4	0	0
12	12	1	5	0	0	0	23	10	8	15	1	7	2
13	13	1	1	31	0	1	61	99	99	12	1	1	1
14	14	5	4	24	3	4	63	3	1	4	0	7	0
15	15	4	5	0	4	3	33	3	12	10	0	8	8
16	16	1	5	0	0	1	36	11	8	14	1	7	1
17	17	7	5	0	98	4	39	3	12	8	0	9	0
18	18	1	1	22	9	0	55	1	10	15	1	1	0
19	19	1	1	32	1	1	55	9	7	16	3	1	0
20	20	1	1	24	2	2	34	4	2	16	3	0	1
21	21	3	1	24	5	2	36	6	3	14	2	3	3
22	22	2	1	23	0	3	44	8	5	18	3	1	1
23	23	5	2	25	2	2	80	5	2	18	4	7	0
24	24	1	5	0	5	1	32	7	5	16	3	3	3
25	25	2	1	32	7	4	37	9	6	5	0	0	0
26	27	1	1	40	5	0	49	9	6	16	3	3	3
27	28	1	3	16	6	2	45	11	9	16	3	1	7
28	29	1	5	0	2	0	36	4	2	18	3	1	4
29	30	2	5	0	5	0	23	3	1	16	1	7	3
30	31	1	5	0	1	0	29	5	2	16	4	3	3
31	32	1	5	0	3	1	30	2	11	14	2	1	0

FIGURE 1.6 SPSS Data Editor Window Showing the Data View

Research Center at the University of Chicago and is used to track a diverse and timely collection of social trends in such areas as multiculturalism, Internet use, religion, and so on. The 1993 GSS was based on a random sample of 1,500 individuals over the age of 17 living in the United States. Since this file is included with SPSS 13.0 and 14.0 for Windows, it is likely to be located in the SPSS folder on your computer or network. If it is, highlight it, and load the file by pointing and clicking on the Open button or by pointing and double-clicking on the file name.

You can select other types of files. By pointing and clicking on the down arrow in the Files of type box, you can list a number of other file types in the large box, including a variety of database and spreadsheet formats. Generally, however, you will open only data or output files. After you have directed SPSS to load a file, note the bottom bar on the SPSS window. This is called the status bar, and it will inform you of the status of the SPSS program during processing. When you are loading data, the SPSS status bar tells you that it is getting the file you have requested. SPSS also tells you the number of cases it is reading from the file as the datafile is loaded into memory. After the datafile is loaded, the SPSS status bar will inform you that the SPSS Processor is ready (see Figure 1.6).

THE PULL-DOWN MENUS

Examine the SPSS Data Editor window, as illustrated in Figure 1.6. There are a number of pull-down menus across the top, and a toolbar with various icons appears below the menu bar. To operate the menus, use the mouse to point and click on the menu. To activate a procedure represented by an icon, point to the icon and click on it. Pull down menus are identical in versions 13.0 and 14.0.

For most of the computer assignments in this book, you will work with the File, Data, and Analyze menus. These menus contain the choices you will use for general operations (calling up or creating a datafile), data definition (describing data you've entered manually or examining the characteristics of variables in preexisting datafiles), and statistical operations (using a particular statistical procedure to analyze your data). The File menu also contains the commands for printing output from your SPSS analysis.

Take some time to familiarize yourself with the characteristics of these pull-down menus. After you single-click and then move the pointer from one selection to the next, a pull-down menu appears under each selection. Also, some menus are connected to submenus. These menus will appear when you point to a procedure that contains multiple subprocedures. There are a great many statistical procedures available in SPSS for Windows. You will use some of these procedures and subprocedures while working on the computer assignments in this book. However, because many of these statistical procedures go beyond the scope of this text and require specialized statistical expertise, you will not use them all.

EXAMINING DATA USING THE Data View

When you complete the data loading process, a screen similar to that shown in Figure 1.6 will appear. This is the SPSS Data Editor. The SPSS Data Editor provides two ways of viewing your data, the Data View and the Variable View. The Data View is the default view and has the appearance of a spreadsheet. The numbers appearing in the Data View are your data. The rows in the spreadsheet correspond to one case in the study (note the numbers in the left-hand column). The columns in the spreadsheet correspond to one variable measured in the study. At the top of each column of numbers is a label (e.g., wrkstat or agewed). These labels are called variable names. As shown in Figure 1.6, pointing at the variable name will cause the variable label (a more comprehensive description of the variable) to appear below the variable name. This feature is useful in attempting to quickly understand value and variable definitions.

One cell in the spreadsheet is framed by double-thick black lines. That cell is called the "active" cell. You can make any cell in the spreadsheet active by moving the frame around with the arrow keys on the keyboard. You can also use the mouse

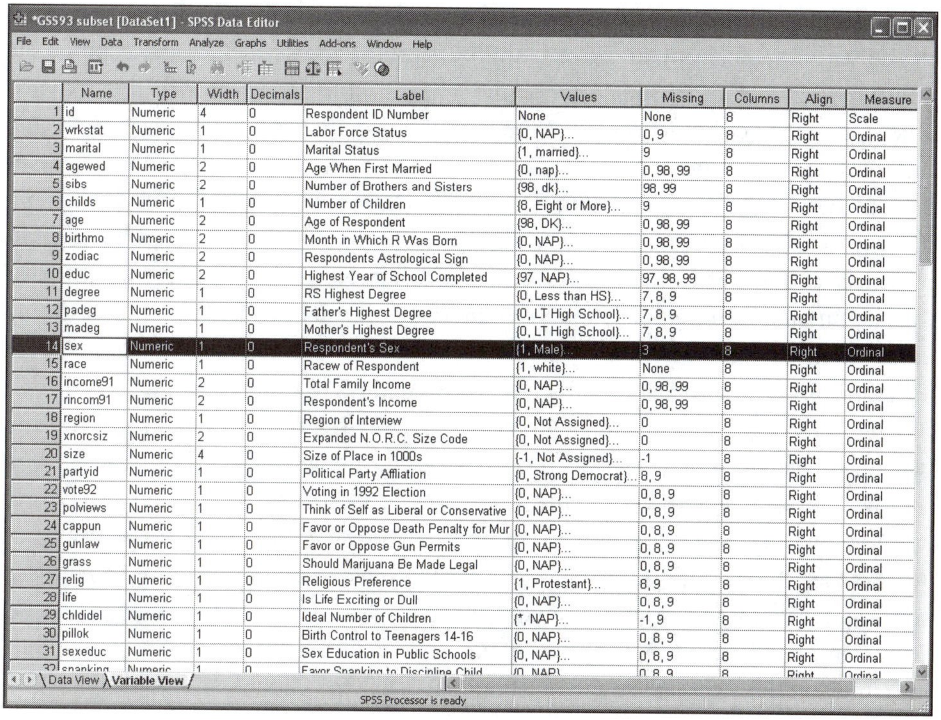

	Name	Type	Width	Decimals	Label	Values	Missing	Columns	Align	Measure
1	id	Numeric	4	0	Respondent ID Number	None	None	8	Right	Scale
2	wrkstat	Numeric	1	0	Labor Force Status	{0, NAP}...	0, 9	8	Right	Ordinal
3	marital	Numeric	1	0	Marital Status	{1, married}...	9	8	Right	Ordinal
4	agewed	Numeric	2	0	Age When First Married	{0, nap}...	0, 98, 99	8	Right	Ordinal
5	sibs	Numeric	2	0	Number of Brothers and Sisters	{98, dk}...	98, 99	8	Right	Ordinal
6	childs	Numeric	1	0	Number of Children	{8, Eight or More}...	9	8	Right	Ordinal
7	age	Numeric	2	0	Age of Respondent	{98, DK}...	0, 98, 99	8	Right	Ordinal
8	birthmo	Numeric	2	0	Month in Which R Was Born	{0, NAP}...	0, 98, 99	8	Right	Ordinal
9	zodiac	Numeric	2	0	Respondents Astrological Sign	{0, NAP}...	0, 98, 99	8	Right	Ordinal
10	educ	Numeric	2	0	Highest Year of School Completed	{97, NAP}...	97, 98, 99	8	Right	Ordinal
11	degree	Numeric	1	0	RS Highest Degree	{0, Less than HS}...	7, 8, 9	8	Right	Ordinal
12	padeg	Numeric	1	0	Father's Highest Degree	{0, LT High School}...	7, 8, 9	8	Right	Ordinal
13	madeg	Numeric	1	0	Mother's Highest Degree	{0, LT High School}...	7, 8, 9	8	Right	Ordinal
14	sex	Numeric	1	0	Respondent's Sex	{1, Male}...	3	8	Right	Ordinal
15	race	Numeric	1	0	Racew of Respondent	{1, white}...	None	8	Right	Ordinal
16	income91	Numeric	2	0	Total Family Income	{0, NAP}...	0, 98, 99	8	Right	Ordinal
17	rincom91	Numeric	2	0	Respondent's Income	{0, NAP}...	0, 98, 99	8	Right	Ordinal
18	region	Numeric	1	0	Region of Interview	{0, Not Assigned}...	0	8	Right	Ordinal
19	xnorcsiz	Numeric	2	0	Expanded N.O.R.C. Size Code	{0, Not Assigned}...	0	8	Right	Ordinal
20	size	Numeric	4	0	Size of Place in 1000s	{-1, Not Assigned}...	-1	8	Right	Ordinal
21	partyid	Numeric	1	0	Political Party Affliation	{0, Strong Democrat}...	8, 9	8	Right	Ordinal
22	vote92	Numeric	1	0	Voting in 1992 Election	{0, NAP}...	0, 8, 9	8	Right	Ordinal
23	polviews	Numeric	1	0	Think of Self as Liberal or Conservative	{0, NAP}...	0, 8, 9	8	Right	Ordinal
24	cappun	Numeric	1	0	Favor or Oppose Death Penalty for Mur	{0, NAP}...	0, 8, 9	8	Right	Ordinal
25	gunlaw	Numeric	1	0	Favor or Oppose Gun Permits	{0, NAP}...	0, 8, 9	8	Right	Ordinal
26	grass	Numeric	1	0	Should Marijuana Be Made Legal	{0, NAP}...	0, 8, 9	8	Right	Ordinal
27	relig	Numeric	1	0	Religious Preference	{1, Protestant}...	8, 9	8	Right	Ordinal
28	life	Numeric	1	0	Is Life Exciting or Dull	{0, NAP}...	0, 8, 9	8	Right	Ordinal
29	chldidel	Numeric	1	0	Ideal Number of Children	{*, NAP}...	-1, 9	8	Right	Ordinal
30	pillok	Numeric	1	0	Birth Control to Teenagers 14-16	{0, NAP}...	0, 8, 9	8	Right	Ordinal
31	sexeduc	Numeric	1	0	Sex Education in Public Schools	{0, NAP}...	0, 8, 9	8	Right	Ordinal
32	spanking	Numeric	1	0	Favor Spanking to Discipline Child	{0, NAP}	0, 8, 9	8	Right	Ordinal

FIGURE 1.7 SPSS Data Editor Window Showing the Variable View

to make a cell in the spreadsheet active by placing the cursor on it and left-clicking on the cell. After making a cell active, avoid making keystrokes on the computer keyboard, because you might accidentally change the value of the cell.

DEFINING DATA USING THE Variable View

Data definition is an important function in the analysis of data. Data definition provides documentation about the data in a file being used or created. Generally, each variable in a datafile has two sets of labels, which describe what the variable is measuring and what values are associated with the measure. In many datafiles, variable and value labels are predefined and stored in the .sav file. In some situations, however, you will want to enter your own data for analysis using SPSS. In this case, you will have to provide a variable and value labels. Regardless of the situation, data definition is an integral component of data analysis. To define your data using SPSS 13.0 and 14.0 for Windows, you will use the Variable View, as shown in Figure 1.7.

The Variable View appears much like the Data View. In the Variable View, however, the variables are listed in rows, and the columns represent attributes of each of the variables. For instance, the first row in Figure 1.7 details the attributes of the sex variable. As you move from left to right across the screen, each of the columns contains an attribute that defines the variable. In the case of the sex variable, the second column notes that the variable is coded as a numeric variable and the third column notes that the variable width is 1 character. The Data View window allows you to perform a number of tasks. If you are using a predefined SPSS datafile, it allows you to examine variable attributes. If you are creating datafiles, it allows you to define your variables. It also allows you to define your scale of measurement as nominal, ordinal, or scale (ratio or interval) and to define missing values and column formats. This book will focus primarily on datafiles with variable and value labels predefined by SPSS.

If you are generating data, however, you will not be using a preexisting datafile. Instead, you will collect your data and create a datafile in SPSS. You might also enter or retrieve data from a word processor such as Microsoft Word, from spreadsheet software like Excel or Lotus, or from a database like dBase or Access (see the Appendix). SPSS has the capability of reading data from all these sources. Or you might prefer to enter your data directly into SPSS using the SPSS Data Editor spreadsheet that appears in the Data View window. When you enter data into the Data View spreadsheet, you should also use the Variable View to define each of your variables.

Familiarize yourself with the Variable View by clicking on each cell in the row for the first case. Buttons appear in all but two of the cells: Name and Label. Whether using a predefined datafile or an original, researchers use both variable names and labels to describe their variables. While the buttons are used to facilitate editing operations in the other cells, you may place the cursor in the Name and Label cells and then create or edit the variable name or label.

In Figure 1.8, the Type column is highlighted. Once it is highlighted, the Variable Type window appears in the foreground. The Variable Type window allows you to define the type of variable used. When using predefined datafiles, however, a researcher will often want to examine how the variables are stored, because SPSS can store them in a number of formats. In most instances, you will encounter variables in predefined datafiles stored as Numeric data such as the sex variable in Figure 1.8. You might also encounter variables stored as String (alphabetic characters including A–Z and 0–9), Date, Dollar, or Custom currency formats.

The Variable Type window also allows you to specify the size of the variable. In the case of the sex variable, Figure 1.8 shows the variable being defined as a Numeric variable with a width of 1 character and 0 decimal places. If your instructor requires you to enter data, you will need to define each of the variables in the datafile you are creating.

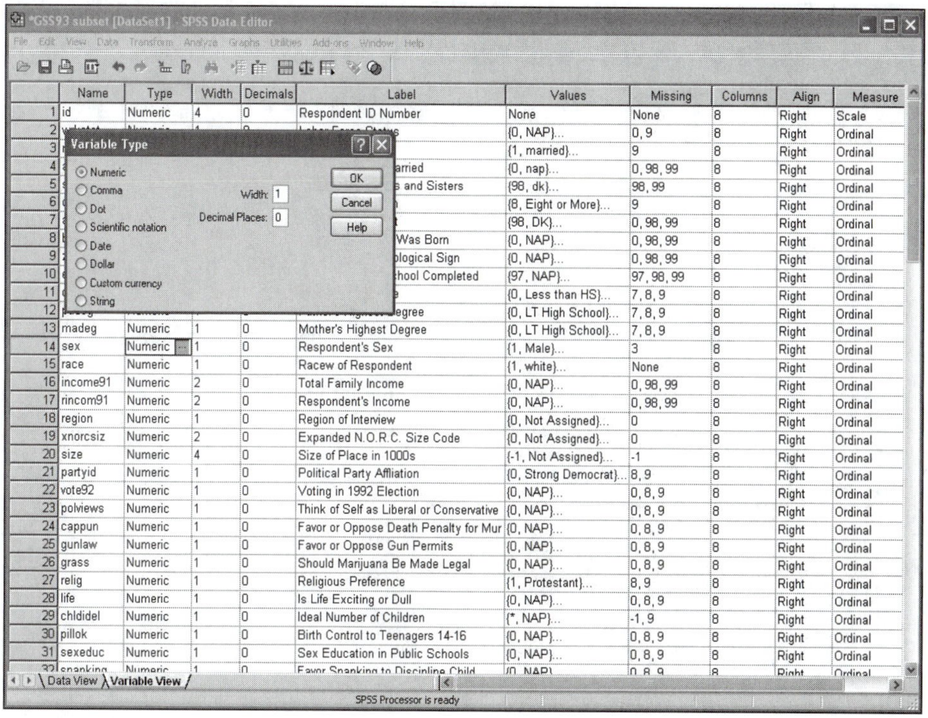

FIGURE 1.8   Defining the Variable Type

Selecting a cell in the Values column will allow you to edit the value labels using the Value Labels window appearing in the foreground. Value labels are used to annotate the values of a particular variable. In the case of the sex variable, Figure 1.9 shows that the value of "1" is associated with the label "Male" and the value of "2" is associated with the label "Female." You may use the Value Labels window to change the labels by clicking on the Change button, to add additional values with associated labels by clicking on the Add button, or to delete a particular label by highlighting it and clicking on the Remove button.

Researchers also must code missing values for cases in which an observation is not recorded for a case. In Figure 1.10, the Missing Values column is highlighted, and the Missing Values window appears in the foreground. In this example, the value of "3" is given to respondents who have missing values on the sex variable. In the Missing Values window you are allowed to insert up to three discrete missing value codes (i.e., "don't know" versus "not applicable"). This allows researchers to discriminate in their coding between different causes for the missing data and allows them to assign values for different types of missing data.

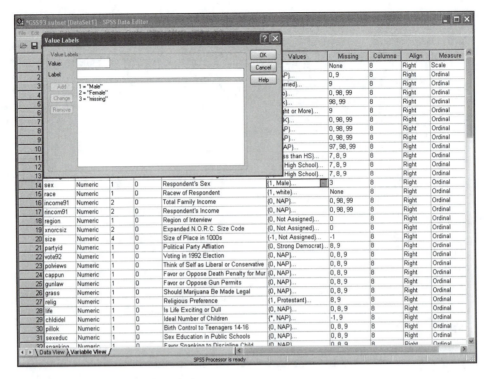

FIGURE 1.9   Defining Value Labels

VIEWING SPSS OUTPUT

SPSS will display the output in the SPSS Viewer screen. Upon completion of SPSS processing, the SPSS Viewer screen automatically appears as the active window on your computer monitor, as shown in Figure 1.11. The SPSS Viewer window has two large display panes. The left-hand pane contains an outline showing the structure of the SPSS output. Specifically, this outline contains information on the type of SPSS procedure performed, the title of the output, notes on the SPSS procedure including program syntax, and the requested statistical output. The right-hand pane contains the output, in other words, tables, graphs, and text produced by SPSS statistical procedures. You can navigate the SPSS output in this pane by pointing and clicking on the element you wish to examine in the left-hand pane. In Figure 1.11, the Statistics portion of the outline is highlighted, and a corresponding arrow with a box is displayed in the right-hand pane. You can move the output on the computer screen by clicking on the arrows that appear on the scrollbars on the right-hand side and at the bottom of the Output screen. These arrows will reposition the output in both panes so that you can examine the entire contents of either pane. At any point during the SPSS session, you can also examine your SPSS output by highlighting SPSS Viewer

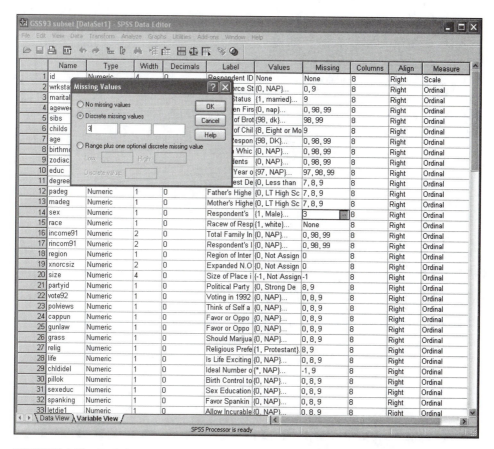

FIGURE 1.10 Defining Missing Values

in the Window pull-down menu (i.e., Window>Output1-SPSS Viewer). Please note SPSS version 14.0 will not read .SPO files created by previous versions of SPSS for Windows. SPSS 13.0 will display .SPO files created by previous versions.

The SPSS output contains a number of items. As shown in Figure 1.12, in addition to the results of your requested SPSS procedures, the Notes section of output contains information on SPSS syntax and information on errors. You will find this information helpful in troubleshooting problems you encounter using SPSS.

PRINTING SPSS OUTPUT

After viewing your SPSS output, you can print the output or save it to a file. To obtain a printout of your SPSS output, you will need to access the File pull-down menu. Figure 1.13 shows the location of the Print command in the File menu. Using the File menu, select the File>Print option.

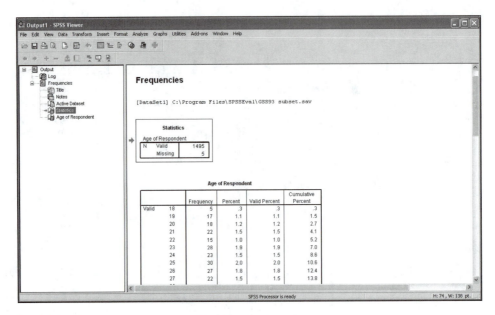

FIGURE 1.11   Active SPSS Viewer Window Showing Output

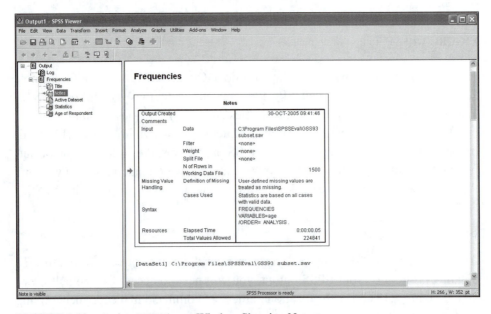

FIGURE 1.12   Active SPSS Viewer Window Showing Notes

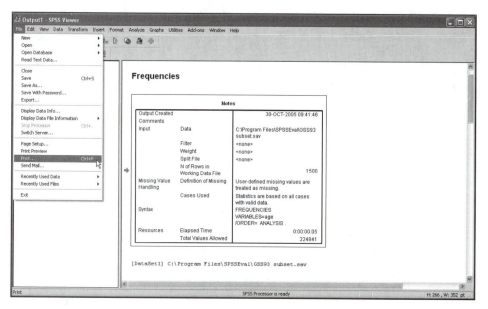

FIGURE 1.13   Selecting the Print Option from the SPSS Viewer Window

After you point and click on the Print option, the Print window will appear, as shown in Figure 1.14. The Print window contains a number of printing options. In the Name box section of the Print window, the default printer device is indicated. If the default printer is the printer you wish to use, you can execute the Print function by pointing and clicking on the OK button at the bottom of the window. This will result in a printout of the SPSS output file on a printer linked to your computer or network. If the output file is large, a few minutes may pass before it begins printing. The Print window allows you several options including choosing a printer, specifying the style of printed output and the number and collation of copies, and printing the output to a file. You can change the printer by pointing and clicking on the down arrow in the box. A list of available printing devices will appear. You can choose another printing device by pointing and clicking on that device.

You can specify the style of printout by pointing and clicking on the Properties button. You can choose among a variety of print options, including portrait or landscape print layouts. Pointing and clicking on the Print to file box will cause a checkmark to appear in the box. After you point and click on the OK button with the checkmark indicated, the Print to file window will appear, allowing you to name the output file and folder for storage. The Number of copies box allows you to specify the desired number of copies either by

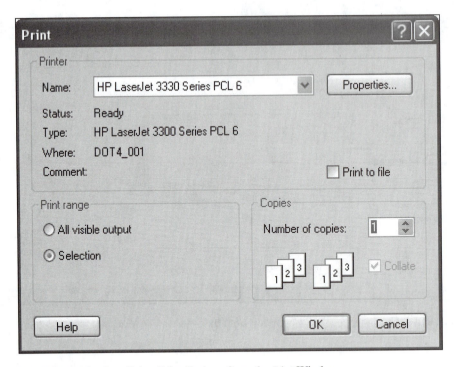

FIGURE 1.14   Specifying Print Options from the Print Window

inserting the number or by increasing or decreasing the value in the box using the up and down arrows. When the value appearing in the box is greater than 1, SPSS allows you to print collated output.

It is very important to make sure the SPSS Viewer window is active when you request a printout of output. If the SPSS Data Editor window is active when you request a printout, you will cause SPSS to print the entire datafile. If you are using a large datafile, you will obtain many pages of useless printout!

EXITING SPSS

If you are using SPSS on a computer used by others (e.g., in a computer laboratory), you should exit the SPSS program as a matter of courtesy. Once you complete your data analysis, access the File pull-down menu and double-click on Exit. SPSS will then prompt you to save your working datafile. If you answer Yes to the query, changes made to the file will be updated and saved. SPSS will also prompt you to save output contained in the SPSS Viewer window to an output file with the filename of your choice. After saving these files, SPSS will terminate and the Windows Desktop screen will reappear.

ON YOUR OWN

You are now familiar with the basics of SPSS for Windows. To complete this assignment, do the following:

1. Load SPSS, and then load a datafile chosen by you or your instructor.

2. Using the Variable View facility, describe two of the variables in your datafile. Describe what they measure and what values are used in the measurement.

3. Take notes on any problems you encountered in accessing SPSS, and report them to your instructor.

ASSIGNMENT 2

Looking at Frequency Distributions and Descriptive Statistics

---

OBJECTIVES

1. Produce a frequency distribution
2. Use SPSS to calculate measures of central tendency
3. Use SPSS to calculate the variance of a distribution
4. Describe the data shown in the SPSS output

This assignment is your first do-it-yourself experience with SPSS. In this assignment, you will learn how to explore your data using the SPSS Frequencies procedure.

Researchers often explore their data before using statistical procedures in the hypothesis testing process. Why would you want to explore your data? The goal is to get a feel for how the data are distributed. You should be interested in the shape of the distribution, the central tendency of the distribution, and the variability of the distribution. Knowledge of these characteristics may help you choose further analysis procedures and guide the interpretation of the results of those analyses.

To explore the data, however, you must first organize the data into a comprehensible form. This allows you to easily and effectively communicate any trends evident in the data. The statistical methods used in organizing and presenting data are often referred to as descriptive statistics. To apply these statistical methods, you will use the Frequencies procedure to produce frequency distributions and calculate measures of central tendency and variance. By using this procedure and SPSS graphics, you will be able to exhibit properties of the distribution using a variety of graphical presentations.

USING THE Frequencies PROCEDURE

You will access the Frequencies procedure after starting SPSS. As described in Assignment 1, start SPSS either by double-clicking on a shortcut icon for SPSS on the Windows Desktop or by pointing to the SPSS icon in the program listings in the Windows Desktop Start menu (i.e., Start>Programs>SPSS for Windows>SPSS

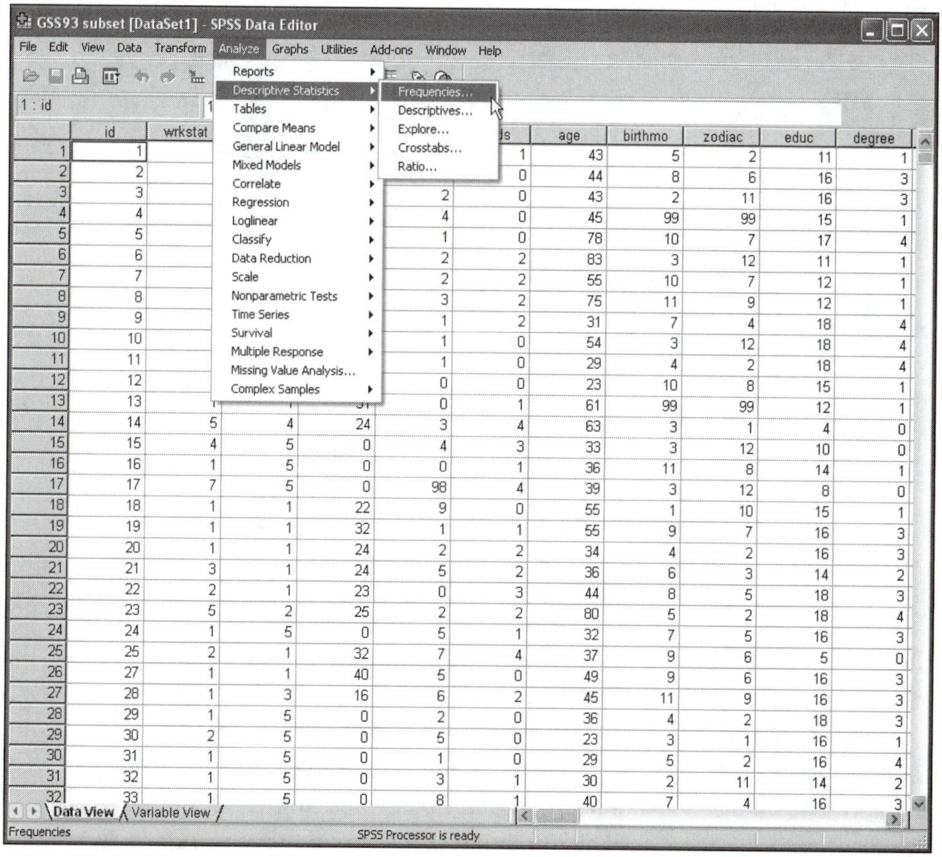

FIGURE 2.1 Choosing the Frequencies Procedure

14.0 for Windows). After you complete this process, SPSS will load, and the SPSS Data Editor window will appear.

The SPSS Data Editor window will appear as a blank spreadsheet. One common mistake students make prior to using an SPSS statistical procedure is failing to load their datafile. Before you apply the Frequencies procedure (or any statistical procedure), you must load your data. Review Assignment 1 for information about this process. In this assignment, the GSS93 subset dataset is used.

Figure 2.1 shows the Analyze menu with all of the options available for statistical analyses. Select Descriptive Statistics from the Analyze menu by pointing and single-clicking on that option. Once you have selected this option, SPSS will display a second set of options, including procedures used to produce frequencies, descriptive statistics, exploratory statistics, crosstabulations of data, and ratio statistics. These are procedures used to produce descriptive statistics. Next, select

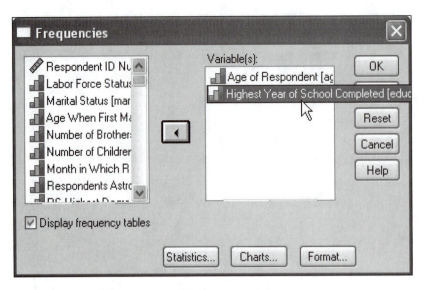

FIGURE 2.2   SPSS Frequencies Window

the Frequencies procedure by pointing and clicking on the Frequencies option on the submenu, as shown in Figure 2.1 (i.e., Analyze>Descriptive Statistics>Frequencies).

After you have selected Frequencies, SPSS will display another screen listing the variables in your datafile, as shown in Figure 2.2. Select the variables you wish to analyze from the variable box in the left-hand side of the window by highlighting them with the mouse and then double-clicking on them. In this example, the variables Age of Respondent (age) and Highest Year of School Completed (educ) were selected. Once you point and double-click on a variable, SPSS transfers it to the right-hand box, as shown in Figure 2.2. The variables in the variable box are listed by their variable labels. The variables can also be listed in the variable box by their variable names (i.e., age and educ). You can shift between these two types of listings by selecting Edit>Options>Output Labels from the main pull-down menu and changing the preference settings. Make sure you consult with your instructor before making such changes.

To produce frequency distributions, you must point and click on the Display frequency tables box. A checkmark will appear in the box, and SPSS will produce frequency distributions for the variable(s) chosen. If you do not want to produce frequency distributions, pointing and clicking on the box will remove the checkmark, and SPSS will not produce frequency distributions.

Functions accessed through the Frequencies window allow you to produce a number of descriptive statistics, including measures of central tendency and graphs. On this screen you can request a chart using the Charts button. SPSS has

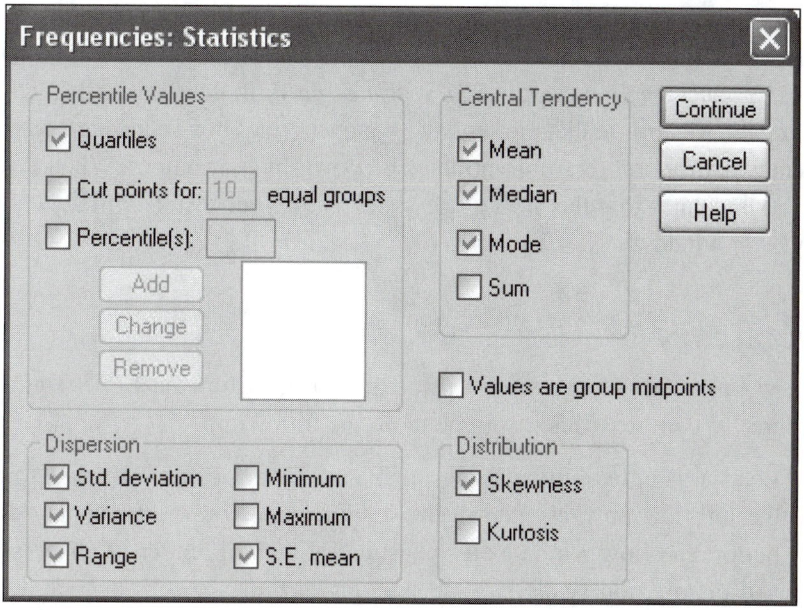

FIGURE 2.3   Frequencies:Statistics Window

powerful graphics capabilities that allow you to present data in graphical form in a number of ways. (Assignment 3 explains in detail how to use this feature.)

Researchers vary in their preferences related to the compilation of frequency distributions. Some prefer to sort the listing of frequencies in ascending order; others prefer descending order. You can click on the Format button in the Frequencies window to change the order in which the frequencies are presented. You can also restrict the number of categories appearing in the distribution for brevity and ease of presentation. Feel free to experiment with the Format options.

You can have SPSS produce descriptive measures for each variable you choose to analyze. When you click on the Statistics button, the Frequencies: Statistics window will appear. You will find a checklist of measures that deal with percentile rank, central tendency, dispersion, and the shape of the distribution. These statistics are primary indicators of the inherent properties of your distributions. You should develop the habit of always specifying some of the Frequencies: Statistics options so that you can examine the characteristics of your distributions. As shown in Figure 2.3, the researcher selected a listing of the quartiles, three measures of central tendency, skewness, and several measures of dispersion. Depending on your instructor's directions, you may choose the same measures for your assignment.

After choosing the appropriate indicators, you must click on the Continue button to return to the Frequencies window. When the Frequencies window reappears, select the OK button. SPSS will then perform the statistical calculations on your data. Make sure that the Display frequency tables box is checked, so that the frequency distribution for your variables is shown in your output. When the SPSS status bar indicates that the SPSS processor is ready, your results will appear in the SPSS Viewer window.

ON YOUR OWN

You now know how to view and print the output from the SPSS Frequencies procedure. To complete this assignment, do the following:

1. Load a datafile, and perform an SPSS Frequencies procedure. Examine the information presented in the output. Focus on the frequency distribution and how it relates to measures of central tendency, dispersion, and distribution shape.

2. In a brief paragraph, describe the characteristics of each of the variables you have analyzed. That is, describe the characteristics of your data using the SPSS output. The best way to accomplish this is by pointing out interesting characteristics about the distributions of your variables. Do not merely report the values and percentages given in the SPSS output, but also write down your thoughts about why these values and percentages are reported (e.g., "Why does variable A have a mean of 40.41?" or "Why is the median for variable A greater than the mean?"). This requires that you study the distributions and measures of central tendency and variability to detect interesting characteristics about the data. In anticipation of describing your data, review the output shown in Figure 1.11 in Assignment 1. The SPSS Viewer window shown in the figure contains information produced by the Frequencies procedure.

ASSIGNMENT 3

Presenting Data in Graphic Form

OBJECTIVES

1. Produce a graph using the Chart option of the Frequencies procedure

2. Obtain a printout of an SPSS graph

3. Describe the meaning of an SPSS graph

In this assignment, you will learn how to produce and interpret a graph in SPSS. The task is not a difficult one. You will start where you ended in Assignment 2.

PRODUCING GRAPHS

When choosing the type of graph to produce, remember to select a graph appropriate for the scale of measure used for the variable(s) you wish to graph. Bar graphs are used to display data measured with nominal scales of measurement. Histograms are used to display variables measured using interval or ratio scales of measure. Histograms are also used to display data from continuous variables.

To produce the graph, you will first need to choose the Analyze>Descriptive Statistics>Frequencies>Charts option from the Frequencies window, as shown in Assignment 2. After you click on the Charts button, SPSS will display the Frequencies:Charts screen, as shown in Figure 3.1. The Frequencies: Charts window allows you to pick either a bar chart, pie chart, or histogram. If you point and click on either Bar charts or Pie charts, the Chart Values area of the Frequencies: Charts window will brighten, allowing you to choose to display either Frequencies or Percentages. If you click on Histograms, SPSS lets you display the normal curve with the histogram by pointing and clicking on the With normal curve box. Proceed with the production of the chart by clicking on the Continue button.

After you click on Continue, SPSS will begin to return to the main Frequencies screen, as illustrated in Assignment 2. After you click on the OK button, SPSS will produce the chart you requested.

After you click on the OK button in the Frequencies window, the chart or histogram will appear with your Frequencies output in the SPSS Viewer window. After you point and click on the named graphics output as it appears in the left-hand pane of the SPSS Viewer window, the graphics you chose to produce will

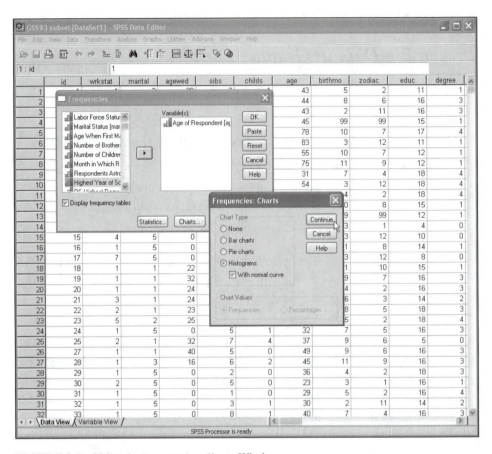

FIGURE 3.1 Using the Frequencies: Charts Window

appear in the right-hand pane. As shown in Figure 3.2, this screen displays the chart that you requested. If you made charts for more than one variable, you can use the arrows that appear in the scrollbars on the right-hand side to position the graphic for viewing. For histograms, SPSS also displays some of the statistics for each variable, such as the standard deviation, the mean, and the number of cases included in the variable. If you choose to display the normal curve, it will be interposed on the graphical display of the distribution.

When you use a histogram to graph interval or ratio data, you should pay particular attention to how the distribution of your chosen variable conforms to the normal distribution. For example, it is easy to see that the distribution of the variable chosen in Figure 3.2 is positively skewed and does not conform to the normal distribution.

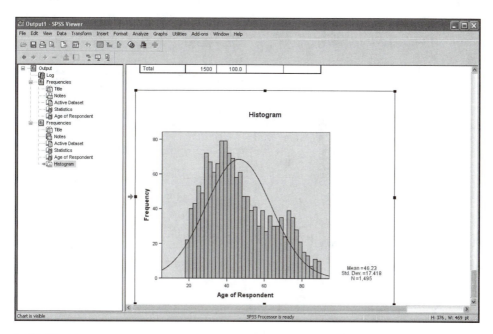

FIGURE 3.2   Displaying Graphical Output in the SPSS Viewer Window

Your final task in this assignment is to obtain a printout of the chart. You will print the graph from the SPSS Viewer window as shown in Assignment 1. SPSS also allows you to export the graph for insertion in other applications. After you point and click on Export (see arrow in Figure 3.2), the Export Output window will appear, as shown in Figure 3.3. This window allows you to choose among a number of exportable graphical file formats for use in other applications (e.g. PowerPoint, HTML). The Export Output window also provides options for adjusting the content and/or quality of the output. You can access these options by pointing and clicking on the Options button. For example, by pointing and clicking on the Options button for an HTML file type, you may access the HTML options, including layering options, chart sizing options, and chart options that allow for adjustments to coloring and compression quality. Options will vary depending on the type of file format exported.

By pointing and clicking on the Chart Size button, you can access options for sizing the chart to fit your needs.

If you choose to print your output using the SPSS Viewer, the computer may take some time to perform this task. Graphical images are memory intensive and require time to load into the printer. In such cases, the computer and printer may appear inactive, and you may need to wait for the image to be processed and sent to the printer.

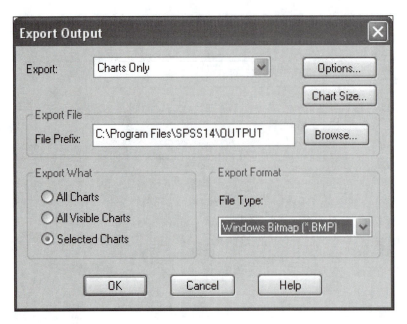

FIGURE 3.3   Export Output Window

ON YOUR OWN

You are now ready to create graphs. To complete this assignment, do the following:

1. Select at least one variable from your data to graph.

2. Obtain a printout of the graph.

3. After printing your chart, in a short paragraph describe each variable you selected. Specifically, describe the shape of the distribution, the skewness, and the relationship of the measures of central tendency to the shape of the distribution. Also, describe the manner in which the data are dispersed. You need to think of the distributions in spatial terms. Be observant and creative in your descriptions!

ASSIGNMENT 4

Testing Research Hypotheses for Two Independent Samples

OBJECTIVES

1. Formulate research questions focusing on differences between two independent populations

2. Use the Independent-Samples T Test procedure to test hypotheses

3. Summarize the results using output from the Independent-Samples T Test procedure

The most common research hypothesis is that two populations are different. To test this type of hypothesis you must choose an appropriate statistical procedure. In this assignment you will learn how to compare interval or ratio data from two independent populations with an independent-samples *t* test.

Prior to comparing populations, however, you must specify the comparisons to be completed. To assist in this, compose research questions to guide the process of hypotheses generation and testing. Research questions are precisely worded queries about what you want to discover by comparing the populations. For example, suppose you want to compare the marital ages of men and women. Implicit in this comparison is a research question that can be formally stated as, "Does the age at time of marriage of men and women differ?"

Based on your observation, you probably have a hunch about the gender differences in marital age. Based on the way the research question is posed, you will need to choose the appropriate statistical procedure to test your hunch. In this example, the research question implies a comparison of males and females. The appropriate test for such a comparison is the independent-samples *t* test.

To properly evaluate the question, you must state these beliefs in the form of testable hypotheses prior to doing the independent-samples *t* test. (Refer to your textbook for more about hypothesis testing.) After stating your hypotheses, you will be ready to test the hypotheses using the independent-samples *t* test.

In this example, the data about the age of men and women when first married comes from the 1993 General Social Survey datafile (GSS93 subset) described in Assignment 1.

FIGURE 4.1   Accessing the Independent-Samples T Test Procedure

The SPSS Data Editor shows the GSS93 subset [DataSet1]. From the **Analyze** menu, the **Compare Means** submenu is open showing: Means..., One-Sample T Test..., Independent-Samples T Test..., Paired-Samples T Test..., One-Way ANOVA...

id	wrkstat						birthmo	zodiac	educ	degree
1						43	5	2	11	1
2						44	8	6	16	3
3						43	2	11	16	3
4						45	99	99	15	1
5				1	0	78	10	7	17	4
6				2	2	83	3	12	11	1
7				2	2	55	10	7	12	1
8				3	2	75	11	9	12	1
9				1	2	31	7	4	18	4
10				1	0	54	3	12	18	4
11				1	0	29	4	2	18	4
12				0	0	23	10	8	15	1
13				0	1	61	99	99	12	1
14	5	4	24	3	4	63	3	1	4	0
15	4	5	0	4	3	33	3	12	10	0
16	1	5	0	0	1	36	11	8	14	1
17	7	5	0	98	4	39	3	12	8	0
18	1	1	22	9	0	55	1	10	15	1
19	1	1	32	1	1	55	9	7	16	3
20	1	1	24	2	2	34	4	2	16	3
21	3	1	24	5	2	36	6	3	14	2
22	2	1	23	0	3	44	8	5	18	3
23	5	2	25	2	2	80	5	2	18	4
24	1	5	0	5	1	32	7	5	16	3
25	2	1	32	7	4	37	9	6	5	0
27	1	1	40	5	0	49	9	6	16	3
28	1	3	16	6	2	45	11	9	16	3
29	1	5	0	2	0	36	4	2	18	3
30	2	5	0	5	0	23	3	1	16	1
31	1	5	0	1	0	29	5	2	16	4
32	1	5	0	3	1	30	2	11	14	2
33	1	5	0	8	1	40	7	4	16	3

PERFORMING THE Independent-Samples T Test PROCEDURE

First, load your datafile. To carry out the Independent-Samples T Test procedure, click on the Analyze pull-down menu, as shown in Figure 4.1. Then click on the Compare Means option. Another submenu will appear listing several procedures by which to compare means, including the One-Sample T Test, the Independent-Samples T Test, the Paired-Samples T Test, and the One-Way ANOVA. In this assignment, you will choose the Independent-Samples T Test procedure from this list by pointing and clicking on that procedure (i.e., Analyze>Compare Means>Independent-Samples T Test). SPSS uses a capital "T" as the symbol for the t statistic in the name of the procedure. This test is referred to in most statistical contexts with an italicized lowercase t as in "t test." In this discussion, the capital "T" notation will be used only when specifically naming the SPSS procedure.

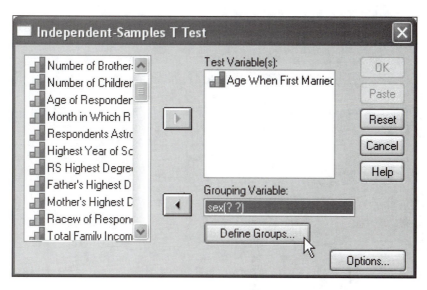

FIGURE 4.2 Choosing Test and Grouping Variables

PICKING TEST AND GROUPING VARIABLES

After you choose the Independent-Samples T Test procedure, SPSS will produce a screen that lists the variables in the datafile. You will choose your variables from this list. You will need to select two things: the Test Variable(s) (dependent variable[s]) and the Grouping Variable (independent variable). The test variable is that variable or measure on which you want to perform the *t* test to test your hypotheses. The grouping variable defines the two independent samples you want to compare using the Independent-Samples T Test procedure.

In the example used in this assignment, you are interested in comparing the marital age of men and women. To accomplish this, choose sex from the list and insert it into the Grouping Variable box, as shown in Figure 4.2. For the Test Variable, you insert the agewed (Age When First Married) variable into the Test Variable box.

For this assignment, you will pick your test variables unless otherwise specified by your instructor. You must remember to use variables measured using the appropriate scale of measurement. You should pick variables measured on either interval or ratio scales (in this case age-in-years is a ratio measure). The independent-samples *t* test is not an appropriate procedure to use with nominal or ordinal scales of measure.

When you insert the Grouping Variable into the box, SPSS will prompt you to define its values (i.e., the two question marks). You need to point and click on

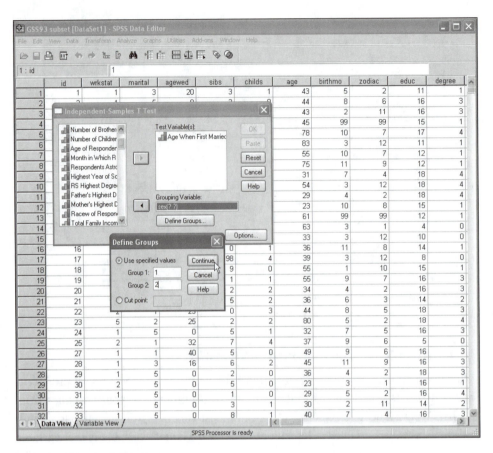

FIGURE 4.3   Defining Groups

the Define Groups button. Another screen will appear allowing you to define the variables, as shown in Figure 4.3. In the GSS example, the sex variable is defined with two values: 1 for "Male" and 2 for "Female." As you can see in Figure 4.3, the value associated with Group 1 is "1" and with Group 2 is "2." Once you have defined the groups, point and click on the Continue button, and you will return to the screen shown in Figure 4.2.

When you complete the selection process, return to the Independent-Samples T Test procedure screen and click on OK to execute the procedure. SPSS will produce the results of the analysis in the SPSS Viewer window. To print the results from the Independent-Samples T Test procedure, invoke the Print procedure when the SPSS Viewer window is active. Review how to print SPSS output from Assignment 1 prior to printing the SPSS Viewer window.

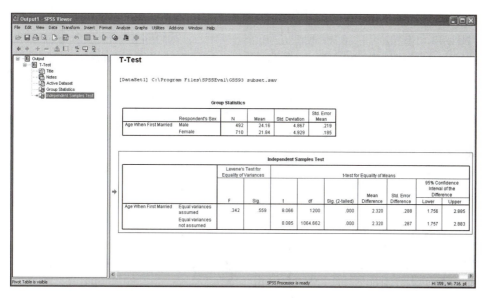

FIGURE 4.4   SPSS Output for the Independent-Samples T Test

INTERPRETING THE OUTPUT

The example used in this assignment provides an illustration for use in the interpretation of results from the Independent-Samples T Test procedure. Interpretation of this type of *t* test from SPSS output is a two-stage process.

The first stage involves assessing the homogeneity of variance between the populations. When using the independent-samples *t* test to test hypotheses, a number of assumptions about the populations being compared must be made. When using this form of the independent-samples *t* test, the researcher assumes that the variance in the populations being compared is the same. The Independent-Samples T Test procedure tests this assumption by using Levene's Test for Equality of Variances, as shown in Figure 4.4. This test is based on the *F* statistic (something you will learn more about in later assignments). SPSS computes both an *F* value and a *p* value. The *F* value is the value computed for the *F* statistic. The *p* value is the calculated probability for making a Type I error (sometimes referred to as the obtained alpha level). The *p* value allows you to determine whether the populations have equal variances. If the *p* value is less than .05 ($p < .05$), the Levene's test indicates that the variances between the populations are not equal. But if the *p* value is greater than .05 ($p > .05$), the population variances are close.

Based on the results obtained in the first stage of the interpretation process, you can now evaluate the hypothesis tested using the independent-samples *t* test.

The results of the *t* test are found on the printout (and in the SPSS Viewer display as shown in Figure 4.4) as a table summarizing the *t* test for Equality of Means. The table contains two rows of results. One row of results is labeled as Equal variances assumed and the other row as Equal variances not assumed. If the Levene's test indicates that the variances of the two populations are about the same, you should use the results as indicated in the row where variances are assumed to be equal. If the Levene's test indicates unequal population variances, you should use the results as indicated in the row in which equal variances are not assumed.

In the example used in this assignment and illustrated in Figure 4.4, the Levene's test indicates that equal variances are assumed (as evidenced by the *F* value not being significant at $p < .05$). In this case, you should use the results indicated in the Equal variances assumed row. These results are different from those indicated in the other row. In the event of unequal variances in the two populations, SPSS mathematically adjusts the results of the *t* test to account for the inequality.

Once you determine which row of results to use, you can then directly evaluate the results of the Independent-Samples T Test procedure. SPSS calculates a *t* value and provides the degrees of freedom (df) on the printout. These are the same values you would calculate manually using the GSS data. SPSS also calculates a Sig. (2-tailed) value, the actual probability of making a Type I error. If this value is less than the specified alpha level (usually $p = .05$ or $p = .01$) used in testing the set of hypotheses, you should "reject" the null hypothesis. By rejecting the null hypothesis, you would conclude that the two populations are different. By contrast, values greater than the specified alpha level require that you "fail to reject" the null hypothesis. In this case, you would conclude that insufficient evidence exists to suggest that the two populations are different. You should not conclude that the two populations are the same.

In the example used in this assignment, SPSS indicates that the Sig. (2-tailed) value is less than .05, a normally specified alpha level. Consequently, you can conclude that a statistically significant difference exists between the marital ages of men and women. By doing so, you are rejecting the null hypothesis that states that no difference exists between men and women in terms of their age when first married. The means indicated that the average marital age of men is greater than that of women. When significance values drop below .001, SPSS prints ".000." This should be read as "$p < .001$."

ON YOUR OWN

You are now ready to test some of your hypotheses using the Independent-Samples T Test procedure. To complete this assignment, do the following:

1. State your research question based on the two variables you choose or your instructor directs you to use.

2. State your hypotheses for use in comparing two independent samples.

3. Test your hypotheses using the Independent-Samples T Test procedure.

4. Write a one- or two-sentence conclusion detailing the results and the meaning of your hypothesis test in terms of the dependent and independent variables.

ASSIGNMENT 5

Testing Research Hypotheses About Two Related Samples

OBJECTIVES

1. Devise questions about differences between two related populations

2. Formulate hypotheses based on the research questions

3. Use the Paired-Samples T Test procedure to test the stated hypotheses

4. Use output from the Paired-Samples T Test procedure to summarize and interpret the results of the hypothesis test

---

The Paired-Samples T Test procedure is used to evaluate whether the means of two related populations are different. What relationship between the populations is examined depends on what research question is being asked. A common research question is, "After participants in a study receive treatment X, will they behave very differently than they did prior to the treatment?" Because this question calls for the measurement of the responses of the same group of participants before treatment and after treatment, the research design can be labeled as a repeated-measures design, a pretest/posttest design, or a before-and-after design.

SPSS uses a capital "T" as the symbol for the t statistic in the name of this procedure—Paired-Samples T Test. As was explained in Assignment 4, this is not standard practice. In the discussion that follows, the capital "T" notation will be used only when specifically naming the SPSS Paired-Samples T Test procedure. Otherwise, a lowercase italicized t will be used as the symbol for the t statistic.

THE RESEARCH QUESTION AND DESIGN

The Repeated-Measures Design

An example of a repeated-measures study is one that asks, "Does a memory-enhancing drug have an effect on the ability of a group of Alzheimer's patients to memorize a list of words?" Prior to the administration of the drug, memory of the list of words is tested. Following the drug treatment, the memory ability of each patient is measured again. In the *t*-test analysis, the pretreatment score and the posttreatment score are compared for each patient.

The Matched-Groups Design

The paired-samples *t* test is also used when the research question involves differences between two separate groups of participants but the researcher wants to ensure that the two groups are evenly matched before the study begins. The design used to answer this research question is called the matched-groups design. In this design, the researcher matches participants on a variable related to the dependent variable to ensure that relevant characteristics of one group correspond to those of the other group prior to treatment.

An example of a research question that leads to this kind of design is, "Would a group of students that had taken a training program do better on the Graduate Record Examination (GRE) than another group that had not?" If, by chance, one group of students starts out with considerably less academic ability than the other, the study could be seriously biased. To ensure that the bias is minimized, the researcher first determines the grade point average (GPA) of each member of a sample of students and then ranks them according to those averages. The two highest-ranking students are then randomly assigned, one to the control group (no training) and one to the treatment group (receives training). Each successive pair of students in the ranking is assigned in the same way. This ensures that the distribution of academic ability (as measured by each student's GPA) is similar in both groups.

Following the matching procedure, the treatment group receives training while the control group gets none. Both groups then take the GRE. The scores of the members of each matched pair are compared in the *t*-test analysis as if they had been produced by the same person.

THE HYPOTHESES

For both of the research questions, the null hypothesis is that the means of the populations represented by the treatment and control groups are equal (two-tailed test) or that they differ in a direction other than was predicted (one-tailed test). If the null hypothesis can be rejected, then the alternative or research hypothesis can be supported. As in the independent-samples *t* test, the data to be analyzed must be measured on an interval or ratio scale.

The following procedure uses a dataset related to the drug treatment for patients with Alzheimer's disease, as mentioned previously (you can find this

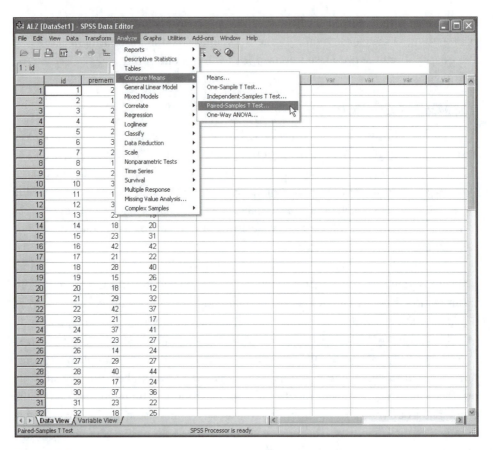

FIGURE 5.1   Selecting the Paired-Samples T Test Procedure

dataset in the Appendix, labeled "ALZ Dataset"). The independent variable is the drug treatment status of the Alzheimer's patient (pretreatment or posttreatment). The dependent variable is the number of words recalled from 100 words presented during a memory test.

EXECUTING THE Paired-Samples T Test PROCEDURE

To perform the Paired-Samples T Test procedure, you must first click on the Analyze pull-down menu. Then, from the Analyze menu, click on the Compare Means option. You will be presented with a final menu listing five statistical procedures used to compare means, as shown in Figure 5.1. Choose the Paired-Samples T Test option (i.e., Analyze>Compare Means>Paired-Samples T Test).

When you choose the Paired-Samples T Test option, SPSS advances to a screen containing a box that lists the variables in the datafile. Because every participant

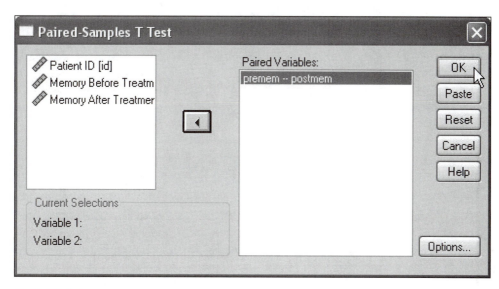

FIGURE 5.2   Selecting the Paired Variables

contributes memory data for both treatment conditions, a grouping variable is not used. Instead, data for both measures of memory ability are entered under two different variable names. In this example, as Figure 5.2 shows, the measures of memory test performance from the pretreatment and posttreatment conditions are represented by the variables premem and postmem, respectively. Click once on each of these two related variables, and then move them into the box labeled Paired Variables by clicking on the arrow button between the two boxes. Run the procedure by clicking the OK button.

INTERPRETING THE OUTPUT

SPSS will output the results of the analysis after processing the data via the SPSS Viewer, as shown in Figure 5.3. The first two tables provided in the output for the analysis contain descriptive statistics for each of the variables. The first table gives the mean, standard deviation and standard error for both variables. The table labeled Paired Samples Correlations gives the bivariate correlation between the two variables and its statistical significance. It is important to look at this correlation because if it is not significant (Sig. > .05), then questions may be raised about the relationship between the two variables and/or the validity of the experimental design. The correlation shown in this example is relatively high ($r = .820$) and is significant at $p < .001$ (as mentioned in Assignment 4, SPSS provides significance levels to only three decimal places; anything less than that is printed as .000).

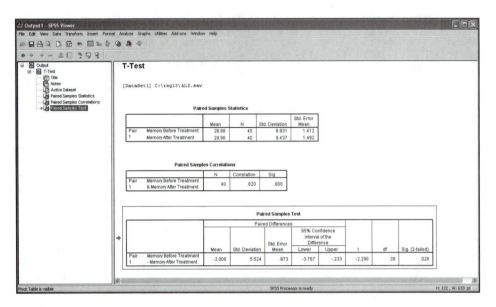

FIGURE 5.3   Output for the Drug/Alzheimer's Example

The Paired Samples Test table reflects the analysis of the difference between the two variables. The most important elements of this group are the *t*-test statistic (*t*), the degrees of freedom (df), and the two-tailed significance level (Sig. (2-tailed)), which appear in the last three columns of the table. These statistics provide the test of the null hypothesis. The *t* value of –2.290 in this example is negative because the mean of the premem scores is smaller than the mean of the postmem scores. The degrees of freedom are equal to the number of pairs of memory test scores minus one (df = 40 – 1 = 39).

The two-tailed significance level for this analysis directly provides the probability for making an error in judgment (Type I error) if you choose to say there is a difference between the populations represented by the two variables. The criterion value (usually $p < .05$ or $p < .01$) for making such errors is compared to the value provided by the *t* test. If the value obtained from the analysis is less than the criterion value you chose before the study began, then you may reject the null hypothesis and infer that the means of the two populations represented by your variables are different.

For the drug and Alzheimer's example, a criterion value for making an alpha error of $p < .05$ and a two-tailed hypothesis are assumed. The two-tailed probability value obtained from the analysis (Sig. = .028) is less than the criterion value. If you have a one-tailed hypothesis, divide the two-tailed significance level in half ($.028 \div 2 = .014$), and make the comparison with your criterion value. In either case, you would be correct to reject the null hypothesis and support the notion that the drug significantly affected the memory of Alzheimer's patients.

ON YOUR OWN

You are now familiar with the use of the Paired-Samples T Test procedure to examine pairs of related variables. Using your data or a dataset provided by your instructor, analyze the effect of the independent variable on the dependent variable. To complete this assignment, do the following:

1. Write out (in words) the null hypothesis and an alternatative (research) hypothesis for this analysis. Specify whether it is a one-tailed or two-tailed hypothesis.

2. Perform the analysis using the Paired-Samples T Test procedure, and print the output.

3. Write a one- or two-sentence conclusion detailing the results and the meaning of your hypothesis test in terms of the independent and dependent variables.

ASSIGNMENT 6

Comparing Independent Samples with One-Way ANOVA

OBJECTIVES

1. Formulate a research question about the differences among more than two groups

2. Test a hypothesis about multiple groups using the SPSS One-Way ANOVA procedure

3. Provide an interpretation of the results of the One-Way ANOVA procedure

4. Provide an interpretation of the Tukey and Scheffé post hoc tests

Many analytic scenarios require the researcher to compare more than two populations or treatment conditions. Suppose you are a biomedical researcher studying the effects of new pharmaceuticals on headaches. You might want to compare the headache relief effectiveness of several different types of drugs. Similarly, you might want to assess the differences among individuals given a range of dosages of the same drug.

Making simple comparisons between two populations or treatment conditions is not problematic because *t* tests provide an appropriate statistical test for such comparisons. Making comparisons of multiple treatments or populations, however, complicates the process. You would need to do multiple *t* tests to complete all of the needed comparisons. The one-way Analysis of Variance (ANOVA) statistic provides a means for making statistical comparisons across more than two groups and alleviates the need to conduct multiple *t* tests. The SPSS One-Way ANOVA procedure performs such comparisons across multiple populations or treatment conditions.

PERFORMING THE One-Way ANOVA PROCEDURE

Suppose you are a cardiologist studying a sample of heart patients. One of the dependent variables you could use in the study is the cholesterol count of your patients. You also know the age of each of your patients, so an interesting research question might be, "Does the cholesterol level of the patients differ significantly across the young, middle-age, and older patients?"

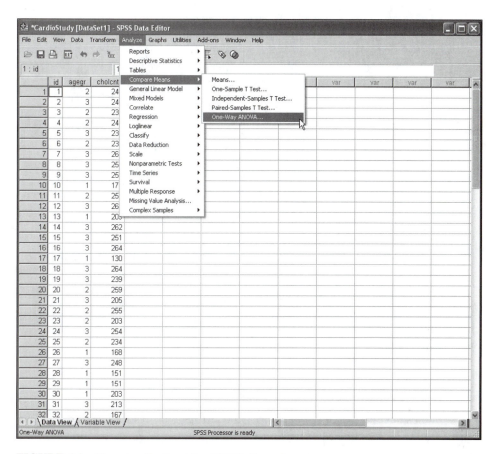

FIGURE 6.1   Choosing the One-Way ANOVA Procedure

To answer this question, you might divide patients into three age-defined groups: those under 35 years of age (young), those between 35 and 50 years old (middle-age), and those over 50 (older). You would then examine the mean cholesterol levels for each group.

Given the way the research question is phrased and that you have more than two groups to compare, the one-way ANOVA is the appropriate statistical test for this analytic scenario. To conduct the test, you would record the ages and cholesterol levels of your patients, load these data into SPSS, and access the One-Way ANOVA procedure. You can find the dataset used in this assignment (labeled "CardioStudy") in the Appendix.

As shown in Figure 6.1, you first move the mouse to the Analyze pull-down menu and highlight the Compare Means procedure. A submenu will appear

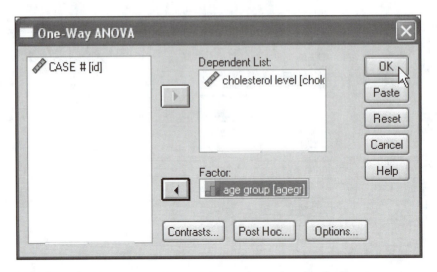

FIGURE 6.2   Moving Variables to the Dependent List and Factor Boxes

containing different procedures used to compare means. Click on One-Way ANOVA, and SPSS will display the One-Way ANOVA window.

CHOOSING AND DEFINING YOUR VARIABLES

The variables in your datafile will appear in the left-hand box of the One-Way ANOVA window, as shown in Figure 6.2. Choose your dependent variable by clicking on it and then clicking on the arrow pointing at the Dependent List box. In this assignment, the variable cholcnt (cholesterol level) has been chosen as the dependent variable. Clicking on the variable agegr (age group) and then on the arrow pointing at the Factor box selects it as the grouping variable. The values for the variable agegr are labeled as follows: 1 = "Under 35," 2 = "35 to 50," 3 = "Over 50."

PERFORMING POST HOC TESTS

You can choose to have SPSS do post hoc tests, calculate descriptive statistics, and run homogeneity-of-variance tests by clicking on the Post Hoc and Options buttons in the main One-Way ANOVA window. First, click on the Post Hoc button to access the Post Hoc Multiple Comparisons window, as shown in Figure 6.3. SPSS allows you to select a variety of post hoc tests for assessing the differences between pairs of means. Post hoc tests are similar to *t* tests in that they test the difference between pairs of sample means. (Remember that the one-way ANOVA statistic only tells you whether there is a difference among all the sample means.) You might, as in

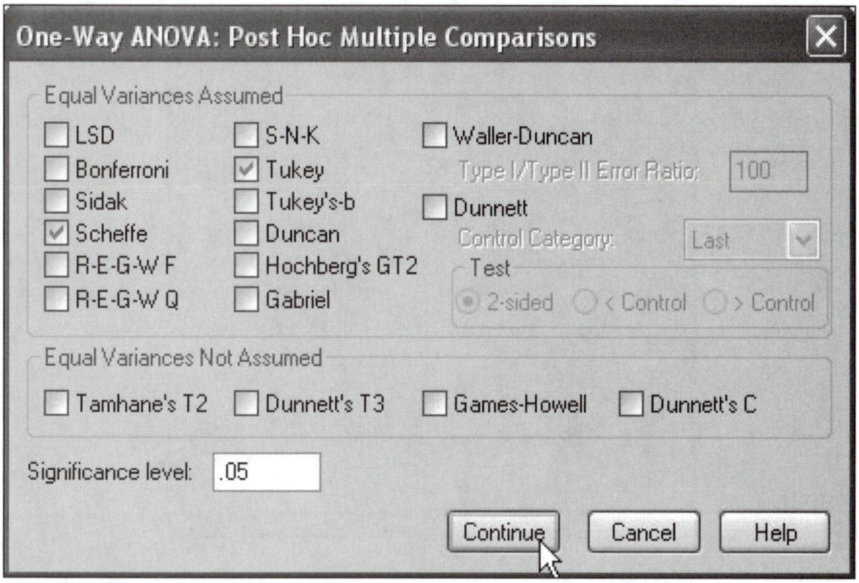

FIGURE 6.3 Choosing Post Hoc Multiple Comparisons

this example, choose Tukey (Tukey's honestly significant difference [HSD] test) and Scheffe (the Scheffé test) by clicking on the name of the test. You will find SPSS output for both of these tests at the end of this assignment. Click the Continue button to return to the One-Way ANOVA window.

To specify the descriptive statistics that can be useful in the overall analysis, click on the Options button. SPSS will present the Options window, as shown in Figure 6.4. Click on the Descriptive box for descriptive statistics and the Homogeneity of variance test box to select Levene's test for homogeneity of variance. After you click on the Continue button, the main One-Way ANOVA window will reappear. You can now instruct SPSS to compute the One-Way ANOVA procedure by clicking on the OK button.

INTERPRETING THE RESULTS FROM THE
One-Way ANOVA PROCEDURE

After you run the One-Way ANOVA procedure, the output from the analysis (shown in Figures 6.5 and 6.6) is presented in the SPSS Viewer window. Figure 6.5 shows the descriptive statistics calculated by the One-Way ANOVA procedure. These statistics include the number of observations in each age group, the mean cholesterol count for each age group, and the standard deviation and standard errors for each group. You will find the confidence intervals for the means as well.

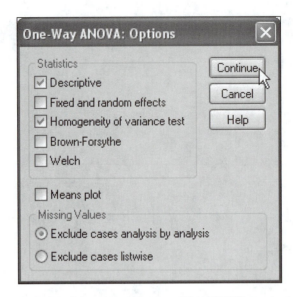

FIGURE 6.4 Options for One-Way ANOVA

The output also shows the result of the test for homogeneity of variance for the age groups. As with the Independent-Samples T Test procedure, SPSS calculates the Levene statistic for this test. In this example, the variances of the three groups do not differ significantly as indicated by the Sig. value greater than .05 ($p = .637$).

A summary of the ANOVA results are shown at the bottom of Figure 6.5. The ANOVA table includes the Between Groups, Within Groups, and Total sums of squares and their corresponding degrees of freedom (df). In addition, the Between Groups and Within Groups mean squares (used to calculate the F statistic) are displayed. However, you should be particularly interested in the F and Sig. columns of the table where the F statistic and the significance level (probability of making a Type I error) are shown.

Figure 6.5 gives the results of the comparisons of the three age groups relative to their cholesterol levels. The Sig. value indicates that the likelihood of committing a Type I error is .000 ($p < .001$). Let's assume you are being conservative and are testing the hypothesis at a criterion alpha level of $p = .01$. Would you be correct to reject the null hypothesis? Certainly, because any probability value less than .001 is also less than a probability value of .01.

INTERPRETING THE POST HOC TESTS

When you examine the group means shown in the Descriptives table at the top of Figure 6.5, you can see differences between the groups being compared in the study. However, the results from the One-Way ANOVA procedure do not indicate

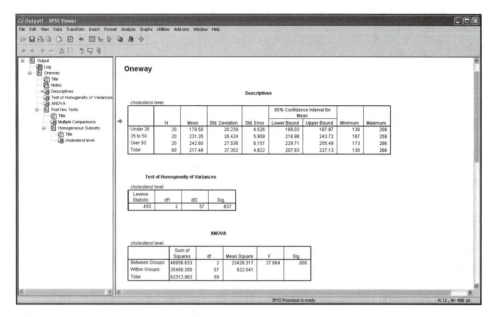

FIGURE 6.5   SPSS Output for the One-Way ANOVA Procedure

which of these group means is statistically different from another. To evaluate where the differences among the age groups lie, you must turn to the results of the post hoc tests calculated by SPSS.

Figure 6.6 shows results from both the Tukey and the Scheffe procedures for each pair of groups studied. In the portion of the Multiple Comparisons table representing the Tukey HSD test, the first and second rows of output show the results of comparing the average cholesterol counts of the Under 35 group to both the 35 to 50 and the Over 50 groups. In the first comparison, between the Under 35 and the 35 to 50 groups, the mean difference is displayed as −52.85*. As noted below the Multiple Comparisons table, the asterisk (*) displayed with the mean difference indicates that a statistically significant difference ($p < .05$) in cholesterol counts exists between the two age groups. Similarly, there is a significant difference (−64.10*) between the Under 35 and the Over 50 groups. However, when you look at the comparison between the 35 to 50 group and the Over 50 group in the second row, you will find that the mean difference between those two groups is not significantly different (no asterisk).

So, in general, the one-way ANOVA indicates a significant difference among your three age groups. However, examining those differences with the Tukey procedure allows you to conclude that this difference was because the cholesterol levels for your younger (Under 35) patients were generally lower than everybody else's. The results of the Scheffe procedure are interpreted in the same manner.

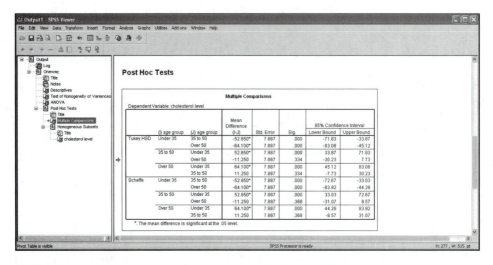

FIGURE 6.6   SPSS Output for the Tukey and Sheffe Post Hoc Procedures

ON YOUR OWN

You are now ready to use the One-Way ANOVA procedure to evaluate the differences between multiple groups. To complete this assignment, do the following:

1. State your research question based on two variables from a dataset that you choose or your instructor directs you to use. Your independent variable should have more than two levels.

2. Test your hypotheses using the One-Way ANOVA procedure.

3. Test between-group differences using a post hoc test.

4. Write a paragraph summarizing the findings as reported by the ANOVA and Multiple Comparisons tables in paragraph form. Describe your results in terms of the relationship between your research hypothesis and the independent and dependent variables.

ASSIGNMENT 7

Comparing Related Samples
with GLM—Repeated Measures ANOVA

OBJECTIVES

1. Devise research questions about differences among three or more populations that are related along the same variable

2. Formulate hypotheses based on the research question

3. Use SPSS and the General Linear Model-Repeated Measures procedure to test the stated hypotheses

4. Use output from the General Linear Model-Repeated Measures procedure to summarize and interpret the results of the hypothesis test

---

A repeated-measures ANOVA can be performed by choosing the Repeated Measures subprocedure of the General Linear Model option in the Analyze menu. The repeated-measures ANOVA is used to analyze results from both repeated-measures and matched-groups designs (see Assignment 5) with three or more treatments representing different levels of a single independent variable. This procedure is commonly referred to as a one-way repeated-measures ANOVA.

Like the paired-samples *t* test, the repeated-measures ANOVA is used to evaluate whether related populations are different. Unlike the paired-samples *t* test, however, the repeated-measures ANOVA can be used to examine the relationships among more than two related populations. The nature of the relationships depends on what research questions are asked and what research designs are used to answer them.

THE RESEARCH QUESTION AND DESIGN

Suppose you observe that infants attend to pictures of human faces longer than they attend to pictures of plain geometric shapes. You could ask, "Does the organization of facial elements (position of eyes, nose, mouth, etc.) have an effect on the ability of infants to perceive human faces?" A simple one-way repeated-measures research design might compare infant attention span in three different conditions to answer this question. A control or baseline condition (Condition A) would involve exposing the infants to a blank, face-shaped oval. In two other conditions, the infants would be exposed to the same oval but with human eyes, nose, mouth, and ears added. In one of these conditions (Condition B), the features would be randomly placed on the oval; in the other (Condition C), the features would be arranged in their typical facial location.

As the phrase "repeated measures" implies, each infant would be exposed to all three conditions. In each condition, the appropriate face stimulus (A, B, or C) would be placed in front of the infant and remain there until the infant looked at it. The length of time (in 0.1-second units) that each infant initially gazed at the picture before looking away would be measured. This measure of attention span would be used as the dependent variable. To control for the effects of experience and order of presentation, each infant would be randomly assigned to a different order of conditions (ABC, ACB, BAC, BCA, CAB, CBA). Because there are six possible orders of the three conditions, an attempt would be made to assign an equal number of infants to each of the six orders.

THE HYPOTHESES

The null hypothesis for the research question tested with one-way repeated-measures ANOVA is that the mean attention spans of the populations represented by the three conditions (A, B, and C) are the same. The corresponding alternative hypothesis is that attention spans are not equal among these populations. Following the initial analysis of the simple hypotheses, post hoc comparisons may be used to investigate more complex issues such as where the significant differences among the conditions might exist.

PERFORMING THE General Linear Model–Repeated Measures PROCEDURE

The following procedure uses a dataset related to the face recognition study previously described. This dataset can be found in the Appendix and has the name "BABE." As in the paired-measures t test, data for the measures of attention span are entered under different variable names. In this example, the measures of attention span from Condition A (Blank Oval), Condition B (Random Features), and Condition C (Correct Features) are represented by the variable names oval, rand, and face, respectively.

To perform a one-way repeated-measures ANOVA, you must first click on the Analyze pull-down menu. Then, from the Analyze menu, click on General Linear Model, as shown in Figure 7.1. You will be presented with a final menu containing a list of General Linear Model procedures. Choose the Repeated Measures item in this menu. In summary, use the following path to get to the Repeated Measures procedure: Analyze>General Linear Model>Repeated Measures.

When you choose the Repeated Measures option, SPSS advances to a screen containing a box labeled Repeated Measures Define Factor(s), as shown in Figure 7.2.

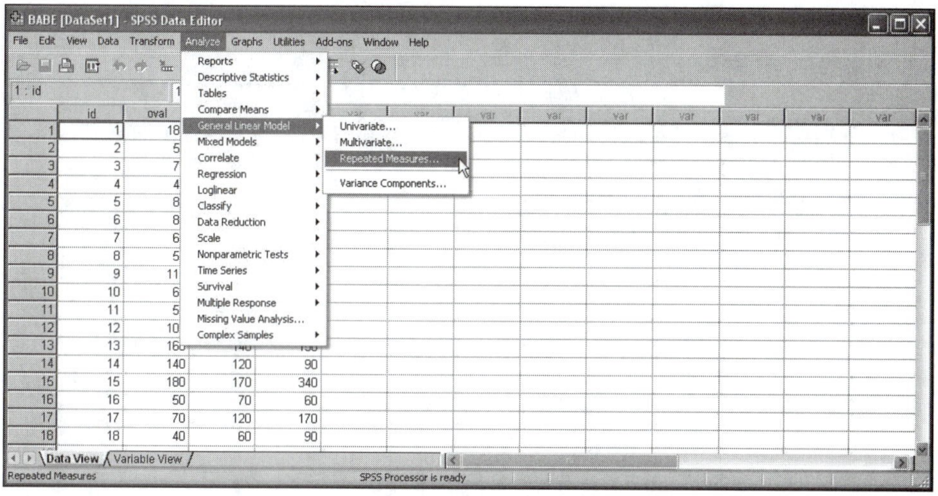

FIGURE 7.1   Selecting the General Linear Model–Repeated Measures Procedure

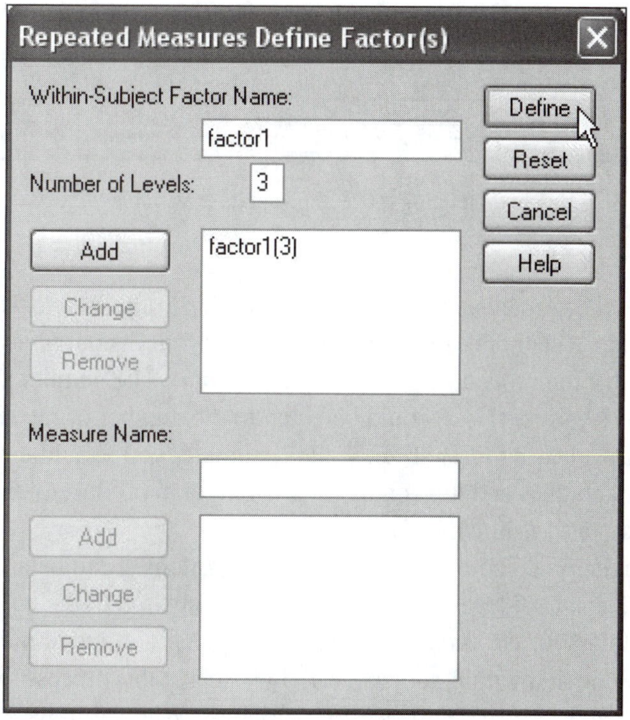

FIGURE 7.2   Indicating the Number of Within-Subject Factors

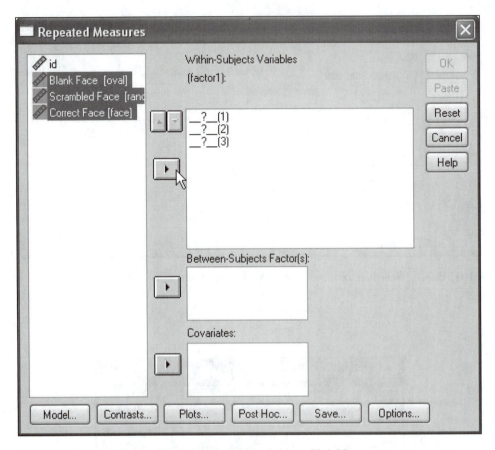

FIGURE 7.3   Selecting and Defining the Within-Subjects Variables

SPSS provides the variable name factor1 in the Within-Subject Factor Name text box. This variable name cannot be one you have used for your data. It is easiest to use the default name that the procedure provides (i.e., factor1). The number you provide in the Number of Levels text box should be equal to the number of repeated-measures conditions that you want to analyze. In this example, you should enter "3." Next, click on the Add button and then on the Define button to produce the screen labeled Repeated Measures, as shown in Figure 7.3.

   In this screen, you should indicate which conditions are to be analyzed. To do this, click on Blank Oval (oval) and then on the arrow button to the left of and pointing to the Within-Subjects Variables (factor1) box. This will move the oval variable into the box as the first within-subjects variable. Repeat this process for the Random Features (rand) and Correct Features (face) variables. When all three variables have been defined (moved) in this way, click on the OK button to run the procedure.

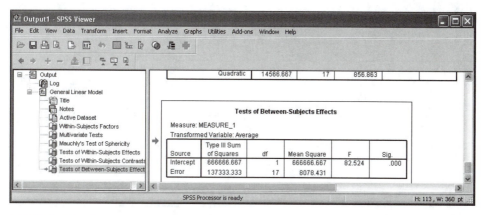

FIGURE 7.4 Testing for Between-Subjects Effects

INTERPRETING THE OUTPUT

Testing Between-Subjects Effects

Figure 7.4 shows the first group of test statistics you should look at in the output from the repeated-measures ANOVA as shown by the SPSS Viewer. This test, labeled Tests of Between-Subjects Effects, examines a special null hypothesis for the analysis. This null hypothesis states that the average of the means of the populations represented by all three variables is equal to zero. In the example used in this assignment, the experimental design allows the researcher to ignore this test. Because timing would not begin until the infants looked at the face stimulus, none of the attention times would be zero. Therefore, this special null hypothesis would be false before any data were gathered. The F value of 82.524 is significant at Sig. $= .000$ ($p < .001$), indicating that the null hypothesis can be rejected. However, if the dependent variable had been defined as difference scores between a baseline condition and the three attention conditions, then it would be possible for the average means for the groups to be zero, and so the Tests of Between-Subjects Effects output would be important for your analysis.

Testing Sphericity

The output for Mauchly's Test of Sphericity is shown in Figure 7.5. It is used like the correlation statistic for the paired-samples t test discussed in Assignment 5. It tests another special null hypothesis concerning an assumption of sphericity that must be met to decide which ANOVA test should be used. Essentially, the sphericity assumption is that the correlations among the attention times for the

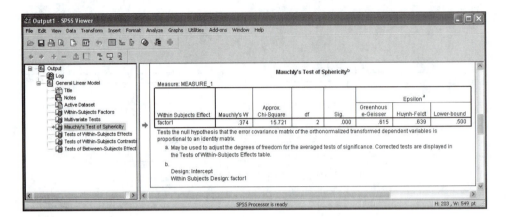

FIGURE 7.5   Testing the Sphericity Assumption

three conditions are equal. The result of the Mauchly sphericity test determines whether a univariate or a multivariate F test should be used for the one-way repeated-measures ANOVA. The chi-square statistic is used to test whether that null hypothesis should be rejected. If the chi-square statistic is not significant (not less than the criterion alpha level decided on before the experiment began [e.g., $p = .05$]), then the univariate test should be used. If the chi-square is significant ($p < .05$), then SPSS provides two alternatives: (1) Correct the test statistic by using one of the epsilon weights (which follow the Mauchly sphericity test) to adjust degrees of freedom for the univariate test, or (2) use a multivariate test. The procedure used in the application of the epsilon weights is fairly complex, so we recommend that you use one of the multivariate tests resistant to violations of the sphericity assumption.

The Multivariate Test

For the attention span example, the sphericity assumption has been violated (Approx. Chi-Square = 15.721, Sig. = .000, [$p < .001$]), so you should use one of the multivariate tests shown in Figure 7.6. The most frequently used statistic is the Wilks' Lambda. In this example, the Wilks' Lambda value of 10.413 with 2 and 16 degrees of freedom is used to test the general null hypothesis that the means of the populations represented by Conditions A, B, and C are equal. The F value used to test this hypothesis is significant at Sig. = .001, indicating that you can reject the general null hypothesis and support the research hypothesis that the arrangement of the features has a significant effect on the attention span of the infants.

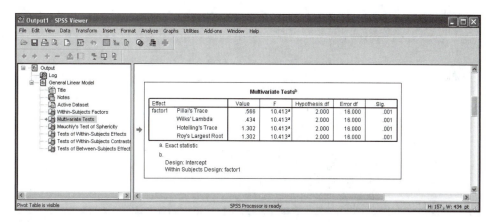

FIGURE 7.6   Multivariate *F* Tests for the Attention Span Example

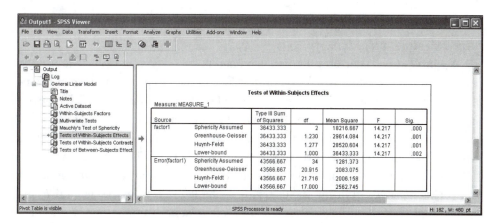

FIGURE 7.7   Univariate *F* Test for the Attention Span Example

The Univariate Test

If the Mauchly sphericity test had not been significant (Sig. > .05), then you could have used the univariate test shown in Figure 7.7. The *F* value for FACTOR1 represents the test of the general null hypothesis. The *F* value of 14.217 with 2 and 34 degrees of freedom would have been significant at Sig. = .000 ($p < .001$).

Post Hoc Comparisons

When the general null hypothesis has been rejected, it is usually of interest to ask which populations differ significantly from one another. If you decide which groups to compare after the data are analyzed, then they are called "post hoc"

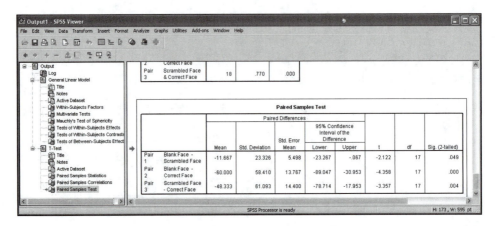

FIGURE 7.8   Post Hoc Comparison *t* Tests for the Attention Span Example

comparisons. Unfortunately, post hoc comparisons are not provided by the Repeated Measures procedure.

A simple way to perform both planned and post hoc comparisons is to use the paired-samples *t* test to make each comparison. In the infant attention span example, there are three possible comparisons: A and B, A and C, and B and C. However, the probability of making an alpha error is not unique to each comparison but is additive across all possible comparisons. In the present example, you would not be testing at the criterion alpha level of $p = .05$, but at $p = .15$ (i.e., .05 + .05 + .05).

The solution to this problem is to make a correction in the criterion alpha level. This involves dividing the criterion alpha level used in the general *F* test by the number of possible comparisons. In the present example, that would be $p = .05 \div 3$ or $p = .0167$. The criterion value of $p = .0167$ is then used for making the decision to reject the null hypothesis for the paired-sample *t*-test comparisons. A one-tailed *t* test should be used, so the probability values obtained for the *t* tests should be divided by two before the significance of the comparison is determined.

Figure 7.8 shows the *t* tests for the three comparisons. The interpretation may be summarized as follows:

Comparison of oval and rand

The obtained alpha value of $p = .0245$ (Sig. = .049 ÷ 2) was not less than the criterion value, $p = .0167$. This means that the *t* value of 2.122 was not large enough to reject the null hypothesis. Therefore, you could not say that infants would respond differently to the blank oval than they would to the oval with scrambled features.

Comparison of oval and face

The obtained alpha value of $p = .0005$ (Sig. $= .001 \div 2$) was less than the criterion value, $p = .0167$. This means that the t value of 4.358 was large enough to reject the null hypothesis. Therefore, you could conclude that attention times for infants looking at a blank oval would be significantly less than for those looking at an oval with normally arranged facial features.

Comparison of rand and face

The obtained alpha value of $p = .002$ (Sig. $= .004 \div 2$) was less than the criterion value, $p = .0167$. The t value of 3.357 was large enough to reject the null hypothesis. Therefore, you would be justified in concluding that attention times for infants looking at an oval with scrambled facial features would be significantly less than for those looking at an oval with normally arranged facial features.

GENERAL CONCLUSION

The comparison of attention times for oval and rand indicates that the mere presence of facial features is not sufficient to significantly change infants' attention to facelike stimuli. However, the comparisons of attention times for face with both oval and rand indicate that attention span is significantly increased when facial features are presented in a configuration typical of a human face.

ON YOUR OWN

You are now familiar with the Repeated Measures procedure to analyze three or more levels of an independent variable. Using your data or data provided by your instructor, you are ready to run this one-way repeated-measures ANOVA. To complete this assignment, do the following:

1. Write out (in words) the null hypothesis and an alternate hypothesis for this analysis. Is it a one- or two-tailed hypothesis?

2. Perform the analysis using the Repeated Measures procedure. If there is a significant difference among your groups, perform post hoc comparisons as described in this assignment. Print the output.

3. Write a one- or two-sentence conclusion detailing the results and the meaning of your hypothesis test. Was there a significant difference among your groups? If so, which groups contributed to that difference?

ASSIGNMENT 8

Measuring the Simple Relationship Between Two Variables

OBJECTIVES

1. Formulate a research question focusing on the covariation between two variables

2. Produce a graph describing the relationship between two variables

3. Provide an interpretation of the Pearson correlation coefficient

---

Correlation is a statistical technique that researchers use to explore the relationship between two variables. These variables are usually referred to as variable X and variable Y. The variables may exist naturally in the environment and are not necessarily manipulated by the researcher.

For example, suppose you are a sociologist interested in the relationship between education and income. The implied research question is, "Do more highly educated people really make more money?" Using data collected on a sample of individuals who work in a geographic area of interest, you would use SPSS to calculate a Pearson correlation coefficient (r) between the two variables. By examining the correlation between the two variables, you will learn about the strength of the relationship between the variables and about the direction of that relationship (i.e., positive or negative). As such, the correlation is sometimes referred to as a "measure of association." Values for correlations range from $r = -1.00$ to $r = +1.00$, indicating perfect negative and positive correlations, respectively. A correlation of $r = 0.00$ indicates absolutely no association between the variables. The absolute value of the range from 0 to 1 indicates the strength of the relationship between the two variables you have chosen.

CHOOSING THE Correlate PROCEDURE

After you have loaded your employee data (the datafile EduWage used in this assignment can be found in the Appendix), you can proceed with the selection of a statistical procedure. If you click on the Analyze pull-down menu, you will see a

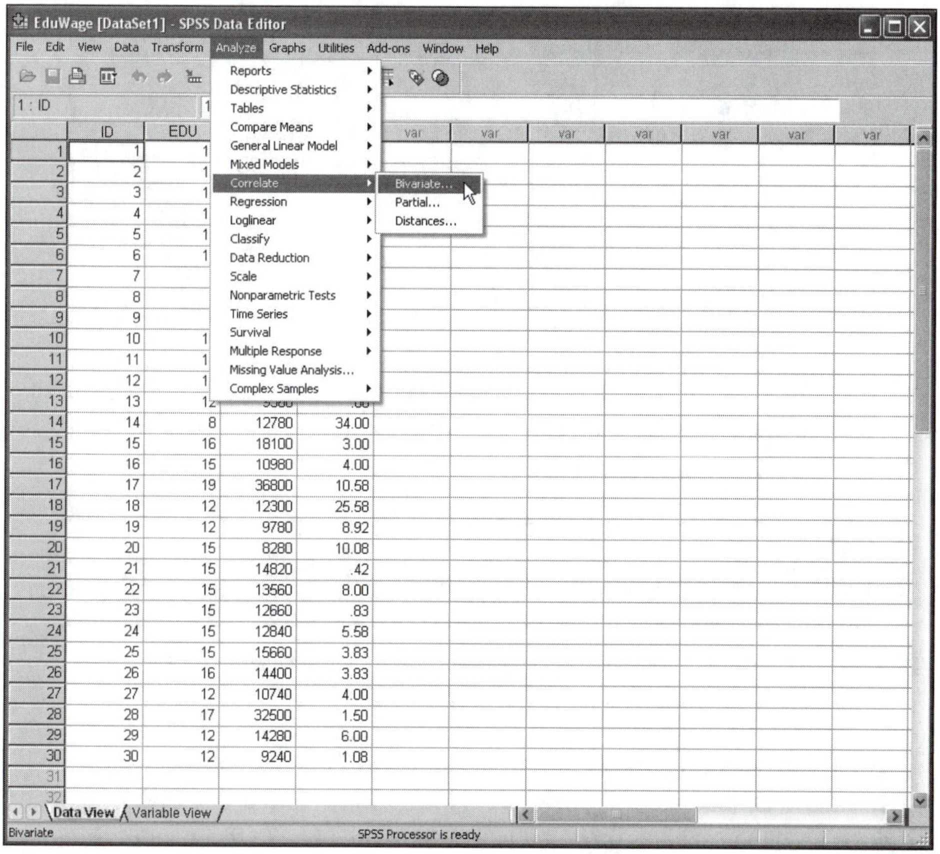

FIGURE 8.1   Accessing the Bivariate Correlation Procedure

menu option labeled Correlate, as shown in Figure 8.1. Highlight the Correlate procedure, and a submenu with another set of choices will pop up. Then click on the Bivariate option. (SPSS will also calculate other forms of correlation such as partial correlations. Consult your textbook or your instructor for more information on these other types of correlational analysis.)

CHOOSING YOUR VARIABLES

In this procedure, you can assess the covariation between any two variables. The Correlate procedure will produce a correlation matrix showing the correlation between all possible pairings of the variables you choose. For example, if you choose two variables, SPSS will produce a 2 X 2 correlation matrix. The size of the matrix is directly related to the number of variables.

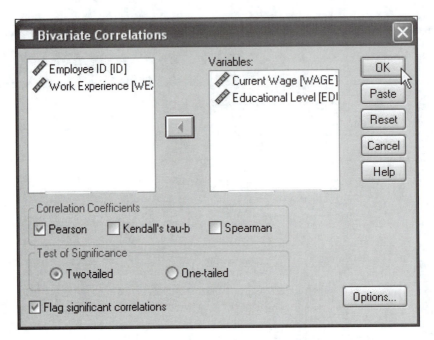

FIGURE 8.2   Selecting Variables for the Bivariate Correlation Procedure

After you select the Bivariate option, SPSS will display the Bivariate Correlations window, as shown in Figure 8.2. This window is similar to variable selection windows used in previous assignments. The left-hand box in this window contains all the variables in the datafile. You will need to choose two variables to complete the analysis. For instance, the example used here assesses the relationship between Current Wage (wage) and Educational Level (edu).

The Bivariate Correlation procedure will produce a correlation matrix showing the correlation between these variables. Once you have chosen your variables and the type of correlation you wish to compute, click on the OK button. SPSS will process the requested statistics.

SPSS offers choices of three different types of correlation for different types of data. The Kendall's tau-b and Spearman procedures are used to calculate correlation coefficients with data measured on nominal or ordinal scales. Only variables measured on either interval or ratio scales are appropriate for use in the calculation of a Pearson correlation coefficient. Both the variables examined in this assignment, edu and wage, are ratio variables; therefore, the Pearson r will be used.

You may also choose to use either a two-tailed or one-tailed test of significance. That choice depends on how you have stated your hypothesis. When you click on

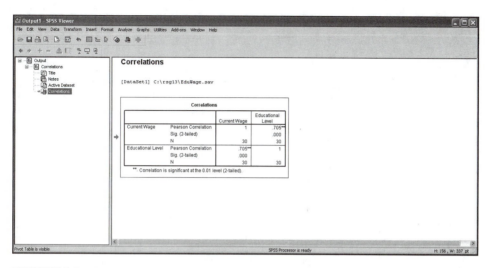

FIGURE 8.3   Output for the Pearson Correlation

the Flag significant correlations box, SPSS calculates *p* values for each Pearson correlation and marks the significant correlations with asterisks (*). As in previous assignments, the *p* value is the actual probability of making a Type I error after rejecting the null hypothesis. In this case the null hypothesis is that the two variables are not correlated. You will reject or fail to reject the null hypothesis based on the criterion level you have stated for testing your hypothesis (usually $p = .05$ or $p = .01$).

Figure 8.3 shows the results of the test for this example. SPSS has calculated a correlation of $r = .705$ between Educational Level and Current Wage among the thirty employees examined. The positive correlation indicates that education and wages vary together in the same direction. In other words, individuals with higher levels of education earn higher wages and those with less education earn lower wages. Note the level of significance indicated on the correlation matrix (*p* value). The results indicate a significant positive correlation between education and income even with a relatively conservative criterion of $p = .01$.

PLOTTING THE DATA

To graphically represent the association between two interval or ratio scale variables, researchers typically plot the data using a scatterplot. The scatterplot will assist you in a number of ways. First, the graphical representation will allow you

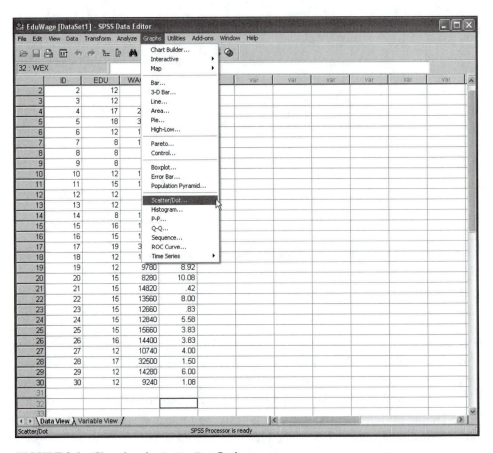

FIGURE 8.4   Choosing the Scatter/Dot Option

to acquire a better feel for how the values of variable *X* covary with values of variable *Y*. You will also be able to detect anomalies that reside in the data. For example, you will easily be able to see points on the graph called "outliers." These are cases that, when plotted on the scatterplot, appear distant from the clustering of most other cases. Outliers may have a profound impact on the correlation you are measuring. Consequently, you might want to identify these as special cases upon which to focus attention and sometimes even eliminate from your analysis.

After you calculate the Pearson correlation, you will produce a graph of the relationship between edu and wage. Use the Graphs pull-down menu from either the SPSS Viewer or the Data Editor and select the Scatter/Dot option, as shown in Figure 8.4.

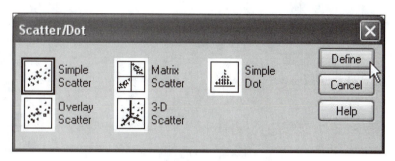

FIGURE 8.5   Choosing the Type of Scatterplot

The Scatter/Dot option is used to represent the relationship between your two variables in the form of a scatterplot. Measures from the two variables are plotted along both the X and Y axes as they occur together in the data. After you choose the Scatter/Dot option, SPSS will produce the Scatter/Dot window, as shown in Figure 8.5.

SPSS allows you to produce different types of scatterplots. You should experiment with the different types of plots to become familiar with the various methods of presenting correlative relationships between your variables. Choose the method of presentation that best describes the attributes of your data. Generally, scatterplots are effective tools for the presentation of data related to the covariation between two variables measured on either interval or ratio scales. Using a scatterplot to present data measured on either nominal or ordinal scales will result in inaccurate presentations of the data. Because you are plotting only two variables on the X and Y axes, you will click on the Simple Scatter option. Then click on the Define button to further define your request.

Next, SPSS will produce the Simple Scatterplot window, as shown in Figure 8.6. Among other things, this window allows you to choose the variables you wish to plot. In this example you would choose wage and edu, the same variables used in the calculation of the Pearson r. Click on the OK button, and SPSS will produce the scatterplot in the right-hand pane of the SPSS Viewer window, as shown in Figure 8.7.

If the scatterplot is satisfactory, you can print it using the Print function in the File pull-down menu in the SPSS Viewer. The processing of the chart for printing takes significant computer resources, so for slower computers, it may take several minutes to print the graphic. Note the Titles and Options buttons shown in Figure 8.6. You can use these functions to embellish your scatterplot for purposes

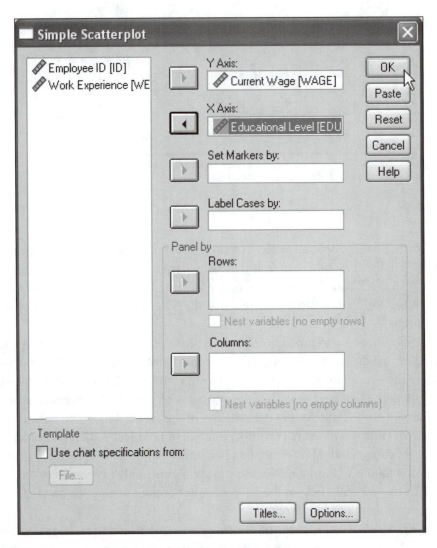

FIGURE 8.6   Choosing Variables for the Simple Scatterplot

of presentation. SPSS will also export graphics to a number of commonly used file formats for storing graphical information.

ON YOUR OWN

You are now ready to explore the relationship between two variables on your own. To complete this assignment, do the following:

1. State your research questions and related hypotheses.

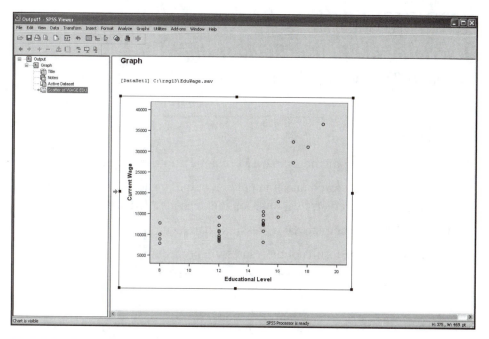

FIGURE 8.7   Output showing the Scatterplot of Educational Level and Current Wage

2. Test your hypotheses using the Correlate procedure.
3. Produce a scatterplot describing the covariation between the two variables you have chosen.
4. Summarize your findings in a brief descriptive paragraph.

ASSIGNMENT 9

Describing the Linear Relationship Between Two Variables

OBJECTIVES

1. Understand the relationship between correlation and regression
2. Formulate a research question predicting the impact of one or more independent/predictor variables on the dependent/criterion variable
3. Use SPSS to compute the regression coefficient(s)
4. Summarize the results of the Regression procedure

In Assignment 8, you learned about measuring the relationship between two variables using the Pearson correlation. You also produced a scatterplot of the X and Y coordinates that described the covariation between the two variables.

In this assignment, you will focus on the description of the relationship between X and Y variables using regression procedures. Regression statistical techniques attempt to find the best way of describing the relationship between a dependent (or criterion) variable and one or more independent (or predictor) variables using a regression line. The regression line represents the "best-fitting" straight line projected through a set of X-Y coordinates like those you produced in the scatterplot in Assignment 8.

Researchers usually assign the independent/predictor variable as the X variable and the dependent/criterion variable as Y variable. As in a correlation analysis, each case in a regression study has a value for both X and Y. When describing the relationship between X and Y, researchers often speak of X as predicting Y. The general mathematical expression for this relationship is as follows:

$$\hat{Y} = a + bX$$

where

$\hat{Y}$ = predicted value for the dependent/criterion variable Y

a = value of the Y intercept (i.e., the point on the Y axis through which the regression line traverses)

b = regression coefficient (i.e., the slope of the regression line)

X = value for the independent/predictor variable X

In this assignment, you will be introduced to SPSS procedures for two forms of regression analysis: bivariate regression and multiple regression. Bivariate regression analysis involves describing the relationship between the dependent/criterion variable Y and one independent/predictor variable X. Multiple regression involves describing the relationship between the dependent/criterion variable Y and more than one independent/predictor variable (X_1, X_2, X_3, etc.). In the multiple regression analysis, you can employ multiple independent/predictor variables to simultaneously predict the dependent/criterion variable. The advantage of multiple regression techniques is that they allow you to predict the dependent/criterion variable in a more comprehensive manner using information from many variables at the same time.

The resulting mathematical expression for the multiple regression technique with two independent/predictor variables is as follows:

$$\hat{Y} = a + b_1X_1 + b_2X_2$$

where

$\hat{Y}$ = predicted value for the dependent/criterion variable Y

a = value of the Y intercept (i.e., the point on the Y axis through which the regression line traverses)

b_1 = regression coefficient for the first independent/predictor variable X_1

X_1 = value for the first independent/predictor variable X_1

b_2 = regression coefficient for the second independent/predictor variable X_2

X_2 = value for the second independent/predictor variable X_2

THE RESEARCH QUESTION

Research questions posed in the context of regression techniques are similar to those posed in the context of the Pearson correlation. In Assignment 8, the research question was, "Do more highly educated people really make more money?" SPSS output suggested a strong positive correlation ($r = .705$) between the two variables. Given this strong correlation, you might ask, "If you know an employee's education level, what is their salary likely to be?" Implicit in this question is the notion that educational level can be used to predict employee salaries.

Bivariate regression analysis is an appropriate statistical technique to use in answering such a question. Furthermore, multiple regression analysis allows you to assess the impact of additional factors as implied by the question, "Does knowing the work experience of the employee, in addition to the education level, "enhance" your ability to predict their current wage?" This question addresses the impact of education on salary while simultaneously recognizing that the work experience of the employee may also be important. In both scenarios, the dependent/criterion measure is the wage earned by the employee. The independent/predictor variables are educational attainment and the employee's work experience.

In the Pearson correlation, the strength and direction of the relationship between X and Y were based on the size and sign of the correlation (r). In a bivariate regression analysis, the impact of the independent/predictor variable(s) on the dependent/criterion variable is assessed in a similar manner using the coefficient of each variable. The larger the coefficient, the larger the effect on the dependent/criterion variable Y in either a positive or negative direction. Independent/predictor variables with coefficients with values near zero have little effect on Y. It is helpful to think of the coefficient as having a multiplier effect on Y. In other words, for every one unit change in X, there is an X times b unit change in Y. Given a value for X, the size of the coefficient b allows you to predict the resulting change in Y.

The goal of multiple regression techniques is to use the multiple independent/predictor variables to explain as much of the variation in the dependent/criterion variable as possible. Ideally, adding more independent/predictor variables to the equation will increase the amount of variation in Y "predicted" by the independent/predictor variables. The amount of variation explained by the independent/predictor variables is known as the coefficient of determination, or r^2. Here, r^2 measures the percentage of variation explained by the independent/predictor variables.

Although you can think of independent/predictor variables within the regression equation as "predicting" the dependent/criterion variable, you must be careful not to assume that a causal linkage occurs between them. In many analytic scenarios, regression analysis represents a more complex form of correlation analysis. Consequently, the presence of a relationship between an independent/predictor variable and a dependent/criterion variable merely denotes association, not necessarily causality.

CHOOSING THE Regression PROCEDURE

The EduWage data used in Assignment 8 are also used for this assignment. After the data are loaded, the first step in performing the analysis is to select the Regression procedure. Once you click on the Analyze pull-down menu, you will

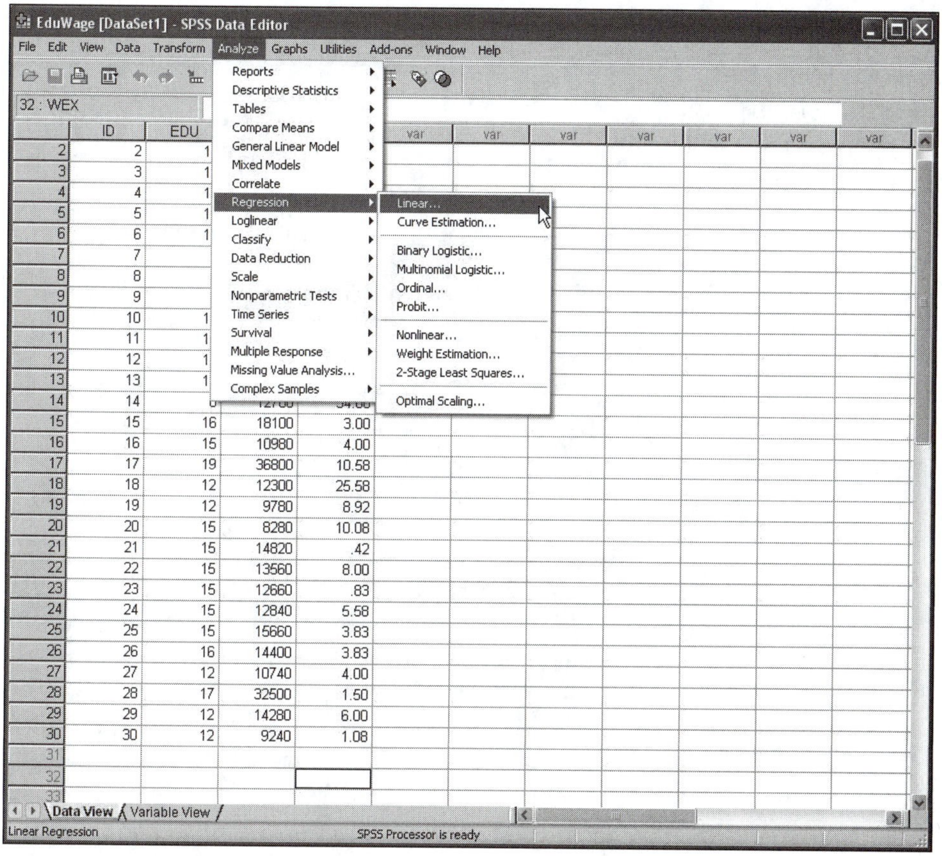

FIGURE 9.1   Choosing the Linear Regression Procedure

see an option on the menu labeled Regression, as shown in Figure 9.1. After you click on Regression, a submenu will appear. You are given several options on this menu.

CHOOSING YOUR VARIABLES

After accessing the Regression menu, you will need to click on the Linear option. SPSS will produce the Linear Regression window, as shown in Figure 9.2. This window allows you to choose both your dependent/criterion and independent/predictor variables. As in previous assignments, click on your chosen variables, and then click on the appropriate arrow button to place the variable in either the Dependent or Independent(s) box. The Dependent box allows you to insert only one variable, whereas the Independent(s) variable box allows you to

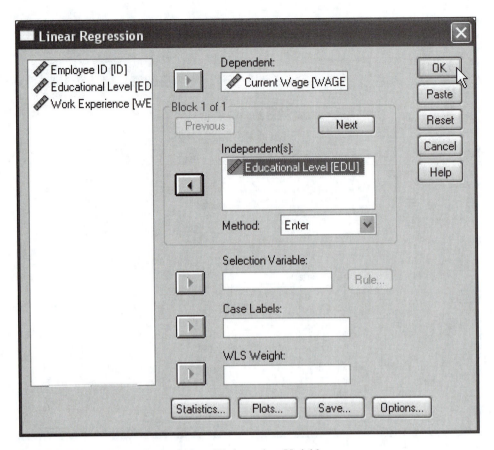

FIGURE 9.2   Selecting Dependent and Independent Variables

insert more than one variable. You will insert one independent/predictor variable for bivariate regression and more than one independent/predictor variable for multiple regression.

Although the Linear Regression window allows you to do several optional tasks related to both basic and advanced regression analysis, you will only deal with some basic descriptive statistics about the variables you are choosing in this assignment. First, click on the Statistics button in the Linear Regression window. SPSS will produce the Linear Regression: Statistics window, as shown in Figure 9.3. You will choose the Estimates, Model fit, and Descriptives statistics from this menu by clicking on each of the choices.

After these options, click on the Continue button, and the Linear Regression window will reappear. Click on the OK button, and SPSS will produce the regression output for this analysis.

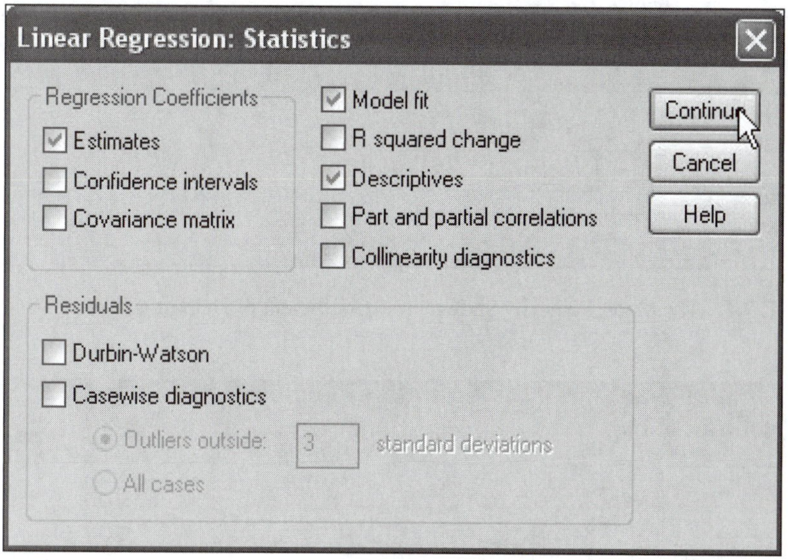

FIGURE 9.3   Choosing Statistics for the Regression Procedure

INTERPRETING THE OUTPUT

In this assignment, two regression analyses are requested. The first analysis is a bivariate regression analysis using wage as the dependent/criterion variable and edu (educational level) as the independent/predictor variable. This type of model represents regression analysis in its simplest form. The second analysis is a multiple regression analysis and involves more than one independent/predictor variable.

In the first analysis, wage is used as the dependent/criterion variable, predicted by the independent/predictor variable edu. The correlation between these two variables was found to be positive in Assignment 8; therefore, you should expect to find a positive regression coefficient for edu.

In the second analysis, you are also interested in the effect that work experience may have on income. To assess the simultaneous impact of education and work experience on wages, the wex (Work Experience) variable is added to the Independent(s) box. SPSS produces output for both of the analyses.

Bivariate Regression

To interpret the bivariate regression analysis you need to focus on three types of information produced by SPSS and shown by the SPSS Viewer. The first item to note is the coefficient of determination displayed in the Model Summary. The coefficient

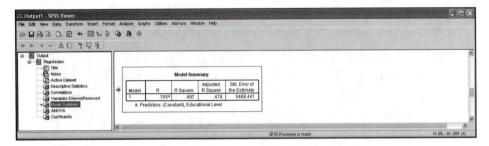

FIGURE 9.4 The Model Summary Showing the Coefficient of Determination

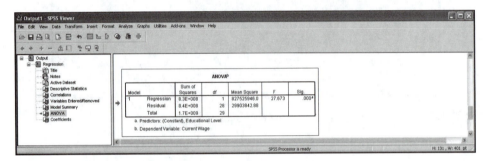

FIGURE 9.5 The F Test for the Significance of the Regression Equation

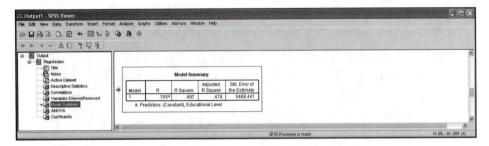

FIGURE 9.6 Significance of the Predictor Variable in a Bivariate Regression Analysis

of determination is designated as R^2 (R Square) as shown in Figure 9.4. In this example, the R^2 value is .497. This indicates that approximately 50 percent of the variation in wages is explained by educational level.

You also need to note the F statistic, shown as $F = 27.673$ in Figure 9.5. If this F statistic is significant at less than the criterion alpha level (usually $p = .05$), you can conclude that the regression equation as computed is statistically significant.

In a regression analysis, however, the most important information is found under the SPSS output heading Coefficients, as shown in Figure 9.6. This section

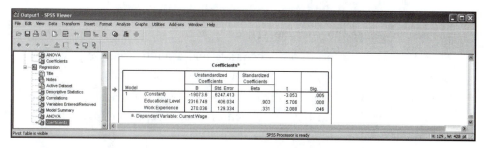

FIGURE 9.7   Significance of the Predictor Variables in a Multiple Regression Analysis

of the output shows you which variables are statistically significant predictors of the dependent/criterion variable. There are three critical components in this section:

B = unstandardized regression coefficient

Beta = standardized coefficient (Beta in this example is .705)

Sig. = computed probability for making a Type I error for the independent/predictor variable edu

In this case, Sig. = .000 ($p < .001$), which is less than $p = .05$ (the criterion alpha level). You can conclude that educational attainment is a significant predictor of salary for the employees, because Sig. is less than the alpha level designated for the analysis.

Multiple Regression

In the multiple regression output, the wex (Work Experience) variable is included in the Coefficients section of the output shown in Figure 9.7. When you examine the output, you will find that previous experience is a significant predictor of salary because Sig. is less than the criterion alpha level of $p = .05$ (i.e., $.046 < .05$). This is a case in which the inclusion of another variable is useful because it provides another explanation for the variation found in the wages of employees and improves prediction. In an analysis of this type, you may want to consider including (as independent/predictor variables) all the variables you think might have an effect on your dependent/ criterion variable.

ON YOUR OWN

You are now ready to use the Regression procedure to further assess patterns of covariation in your data. To complete this assignment, do the following:

1. Formulate a research question, appropriate for use with a regression analysis, related to your dataset.

2. Use SPSS to perform both a bivariate and a multivariate regression analysis using at least two independent/predictor variables.

3. Describe in words the results of both your regression analyses. Be sure to describe the contribution of each of your independent/predictor variables.

ASSIGNMENT 10

Assessing the Association Between Two Categorical Variables

OBJECTIVES

1. Formulate a research question focusing on the association between two variables

2. Use the SPSS Crosstabs procedure to produce contingency tables

3. Use SPSS to compute the chi-square statistic

4. Describe the results of a chi-square analysis

Researchers often face situations in which their data are not measured using interval or ratio scales. They also face situations in which their data are not normally distributed. In such situations, researchers must use a family of statistics referred to as "nonparametric" statistics. Nonparametric statistics differ from parametric statistics such as independent-samples t tests in that nonparametric tests do not require the researcher to make assumptions about the normal distribution of the data and the homogeneity of variance. Given these characteristics, nonparametric tests are well suited for use with data measured on nominal or ordinal scales.

In this assignment, you will begin to use a nonparametric test of difference and independence called the chi-square statistic (χ^2). When using parametric statistics, you are generally dealing with population parameters such as the mean or standard deviation. When using the chi-square statistic, you are interested in looking at differences in the frequency or proportion of events as they occur between two populations. A research question suggesting such a comparison might be, "How does the number of men and women pilots compare to the number of men and women in the general population?" In this situation, you would use the chi-square statistic to assess your observations of the proportion of men and women that you observe to be pilots relative to what you would expect to find based on the proportion of men and women in the general population. Sometimes this type of test is referred to as the chi-square test for goodness-of-fit.

In other situations, you might want to determine whether a relationship exists between two variables measured using a nominal or an ordinal scale. In this analytic scenario, you would use the chi-square test for independence. A research question dictating this type of a test might be, "Does the proportion of male and female pilots vary between airline companies?" This question would require you to examine the proportion of male and female pilots across the major airlines to ascertain whether differences exist in the proportions of male and female pilots, company by company. If you used the chi-square statistic to evaluate this question, finding different proportions in the airline companies would mean that a relationship does exist between gender and airline company. In other words, the proportion of male and female pilots depends on the airline company. Conversely, finding the same proportions across the airline companies would indicate that no relationship exists between the variables.

In this assignment, you will learn how to use the SPSS Crosstabs procedure to calculate the chi-square test for independence. SPSS will construct crosstabulations (i.e., contingency or cross-classification tables) between the classifications (categories) of two variables. The crosstabulation takes the form of a frequency distribution arranged as a matrix in which rows correspond to the categories of one variable and columns correspond to the categories of the second variable. From the intersection of rows and columns emerge cells that represent both.

To obtain the chi-square statistic, SPSS must calculate both expected and observed frequencies as they occur in each cell. Observed frequencies are the actual frequencies that emerge as the result of your cross-classification of the two variables. Expected frequencies are those that the researcher would expect if there were no relationship between the two variables.

To calculate the chi-square statistic, SPSS examines the cumulative magnitude of difference between the expected and observed frequencies in each cell. (Refer to your textbook for the formula for the chi-square statistic.) As the cumulative magnitude of difference between observed and expected cells increases across the cells, the value of the chi-square statistic becomes larger. As the chi-square statistic increases, the likelihood grows that a relationship exists between the two variables.

Prior to performing the Crosstabs procedure, you must specify a research question and state a set of hypotheses to be tested using the chi-square statistic. In this assignment, the GSS93 subset described in Assignment 1 will be used to examine the relationship between gender and views on capital punishment. The implicit research question is, "Do men and women have different views about capital punishment?" To properly evaluate the research question, you must state null and alternate hypotheses. (Refer to your textbook for more about hypothesis testing.) After stating the hypotheses, you will be ready to test them using the chi-square test for independence.

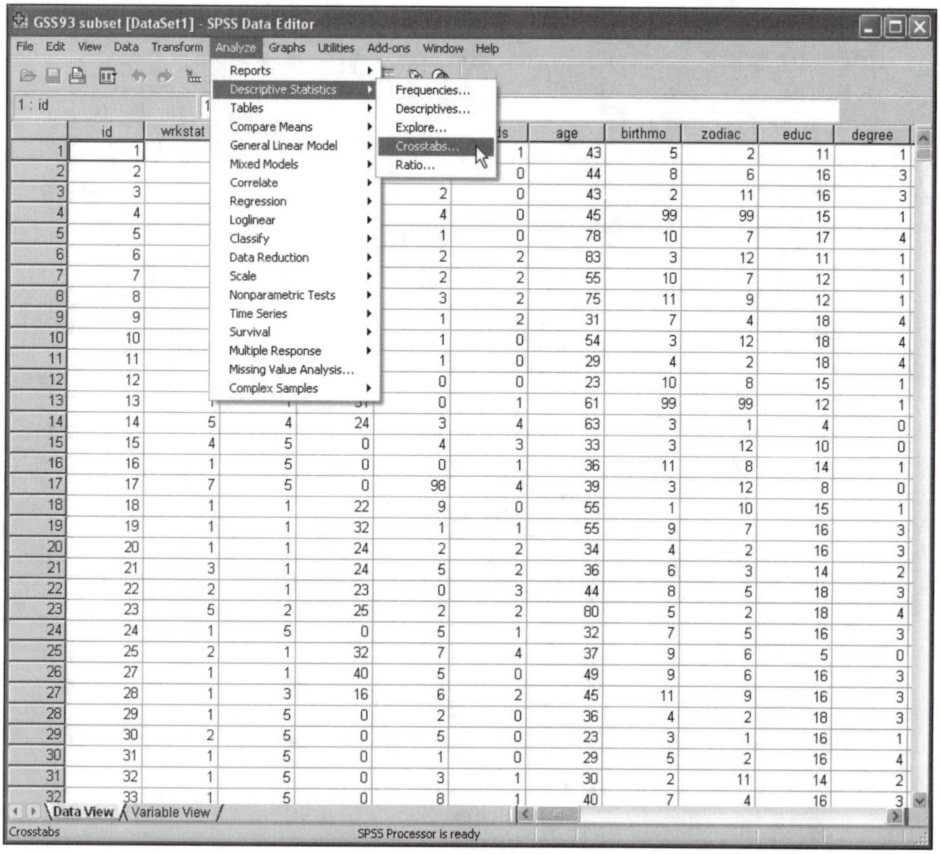

FIGURE 10.1   Choosing the Crosstabs Procedure

PERFORMING THE Crosstabs PROCEDURE

After loading your data, use the Analyze pull-down menu to access the Descriptive Statistics option. Then click on the Crosstabs option (i.e., Analyze>Descriptive Statistics>Crosstabs), as shown in Figure 10.1.

After you access the Crosstabs procedure, SPSS will produce the Crosstabs window. The Crosstabs window lists all variables in the left-hand pane, as shown in Figure 10.2. You will need to move one variable to the Row(s) box in the Crosstabs window and another variable to the Column(s) box. In the present example, the variables of interest are the gender of the respondent (the variable sex) and whether they favor or oppose the death penalty for murder (the variable cappun). Place the sex variable in the row box and the cappun variable in the column box. You can insert a third variable into the Layer 1 of 1 box. Insertion of a variable here

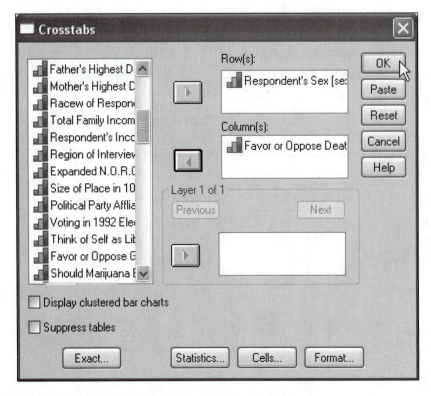

FIGURE 10.2   Choosing Variables for the Crosstabs Procedure

instructs SPSS to produce a three-way contingency table comparing three different categories of variables.

After choosing your variables, you must indicate that the chi-square statistic will be calculated using the Crosstabs procedure. Choose the chi-square statistic by clicking on the Statistics button appearing in the Crosstabs window to produce the Crosstabs: Statistics window shown in Figure 10.3. Although the Crosstabs: Statistics window provides a number of statistical options, only the Chi-square option will be selected for this exercise. After selecting Chi-square, click on the Continue button, and the Crosstabs window will reappear.

To complete the analysis, you also need to click on the Cells button in the main Crosstabs window. SPSS will produce the Crosstabs: Cell Display window with a number of options. To understand the crosstabulation of the two variables, you should select those options shown in Figure 10.4. Choose Observed and Expected under the heading Counts. Then, under the heading Percentages, click on Row and Column. These options will provide the detail needed to understand the chi-square statistic. Once you have chosen the cell attributes, click on the Continue button. The Crosstabs

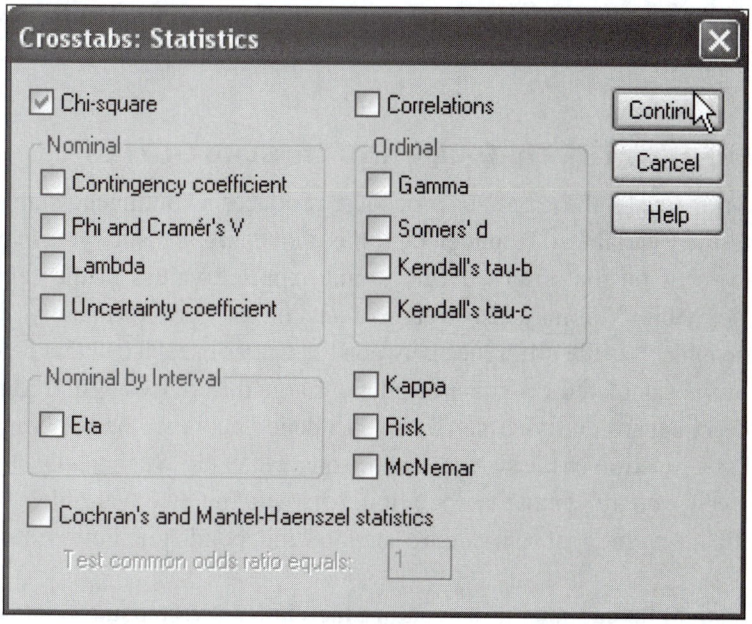

FIGURE 10.3 Selecting the Chi-Square Statistic

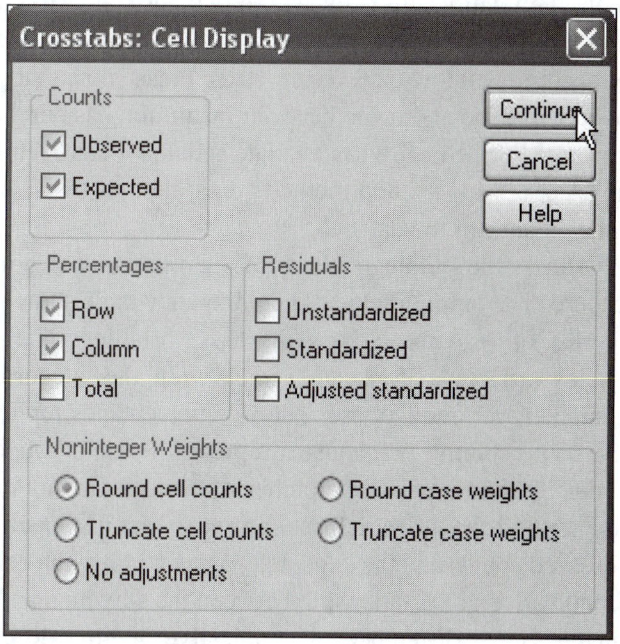

FIGURE 10.4 Choosing Table Attributes

window will reappear. After you click on the OK button in the Crosstabs window, SPSS will produce the contingency table and calculate the chi-square statistic.

INTERPRETING THE Chi-Square AND Crosstabs OUTPUT

As stated previously, the Crosstabs procedure produces a contingency table of two cross-classified variables. To understand the chi-square statistic, you may find it helpful to focus on the observed values and expected values in the cells of the contingency table. You may also want to focus on the "proportionality" between cells in the table. It is the difference between the values in each cell that determines the size of the calculated chi-square statistic. Large differences will produce large values of chi-square and vice versa. Independence between the two variables is assumed as a function of the size of the chi-square statistic. Statistically significant values of chi-square denote association between the two variables, whereas nonsignificant values of chi-square denote independence between the two variables.

You should be cautious in evaluating the chi-square statistic in situations in which cells have fewer than five observations. In such cases, SPSS will give a warning about small cell sizes. The calculation of the chi-square statistic is affected by very small observed frequencies in cells and may provide misleading information about the relationship between the two variables. To compensate for the sparseness of the data in some cells, you may choose the Exact Tests button on the Crosstabs procedure window (see Figure 10.2). Please note that your ability to choose this option will depend on whether your institution has purchased the Exact Tests modules from SPSS. SPSS will calculate reliable exact significance levels based upon the Exact or Monte Carlo methods regardless of the size, distribution, sparseness, or the balance of the data.

Figure 10.5 shows the output in the SPSS Viewer window produced by the Crosstabs procedure. The output includes a contingency table showing the various possible categories of response across the two variables. The categories of response for the sex variable are Male and Female while the categories for the cappun variable include Favor and Oppose. The row percentages represent % within Respondent's Sex. The column percentages represent Favor or Oppose the Death Penalty for Murder. To gain a more comprehensive understanding of the relationship between the sex and cappun variables, however, you must examine the degree to which the observed counts and the expected counts within each cell differ. Large differences are indicative of a relationship between the way men and women view capital punishment. Small differences are indicative of little difference in their sentiments toward capital punishment.

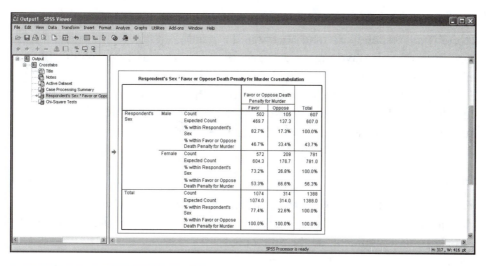

FIGURE 10.5   SPSS Output for the Crosstabs Procedure

FIGURE 10.6   SPSS Output for Chi-Square Tests

Although you may find differences occurring between the observed counts and expected counts, you must examine the chi-square statistic to determine whether an overall statistically significant difference exists. Figure 10.6 shows results from the calculation of the chi-square statistic as it appears in the SPSS Viewer window. The Value column shown in the output is the chi-square value as calculated by SPSS. Similarly, df refers to the degrees of freedom for the chi-square. The calculated p value for the chi-square statistic is shown as the Asymp. Sig. (2-sided) calculation. As in previous assignments, you reject or fail to reject your null hypothesis based on the alpha level you have chosen (usually $p = .05$ or $p = .01$). In this example, the Asymp. Sig. (2-sided) value ($p < .001$) is less than $p = .05$, indicating that men and women have different views on capital punishment. Overall, all respondents tend to favor the death penalty for murder, but the proportion of males favoring the death penalty is larger than the proportion of like-minded females.

Pearson chi-square as shown in Figure 10.6 is not the same as the Pearson correlation. Although Karl Pearson developed both statistics, they address different statistical issues and use different types of data.

ON YOUR OWN

You are now ready to use SPSS to construct contingency tables and to calculate the chi-square statistic to assess the association between two categorical variables. To complete this assignment, do the following:

1. State your research question based on two variables in your datafile.

2. Test your hypotheses using the chi-square statistic.

3. Summarize your findings in a descriptive paragraph about your analysis.

APPENDIX

Entering Data Using Programs Other Than SPSS

USING A TEXT EDITOR TO ENTER DATA

You can use any text editor or word processor to enter and store data. The text dataset is then retrieved and converted to an SPSS datafile by the Text Import Wizard. Data must adhere to one of two specific formats and must be saved in text (ASCII) form. The Text Import Wizard processes the following data formats:

Fixed width

Delimited

These text formats are remnants of the era of punch card and paper tape data entry and have largely been supplanted by more sophisticated data entry approaches including that used by the SPSS Data Editor.

FIXED-WIDTH FORMAT

In fixed-width format, the data for each participant (case) are entered on a single line of text. If there are too many values to enter on a single line, more lines may be allotted to each case. There will always be the same number of lines of text in each case even if they are not needed for some cases.

Each numeric digit, alphabetic character, or symbol that constitutes your variable will occupy a specific column in a particular line. For example, Figure A.1 shows a four-case dataset created using Windows Notepad. The names of the variables are entered on the first line and are separated from one another by a space. Entry of variable names in the text file is optional. If you do not choose to include variable names, the Text Import Wizard will let you do this when the text file is imported, or you may enter variable names and other specifiers once the data appear in the SPSS Data Editor.

In this example, the variables are ID, Name, Age, Gender, and Score. Because the variable names occupy line 1, the first case starts on line 2. The ID number

FIGURE A.1   Four-Case Fixed-Width Dataset Created with Windows Notepad

for the first case, 8953, occupies columns 1–4 in line 1. The ID numbers for the other cases also occupy columns 1–4. The columns containing the values for each variable are separated by a space, so the variables are found in the following columns: ID = 1–4, Name = 6–21, Age = 23–24, Gender = 26, Score = 28–31.

Note the following:

- The Text Import Wizard will recognize Name and Gender as string (text) variables.
- Age is missing for the second case. When a case in fixed-width format has missing values, leave the space in the columns allotted to that variable blank.
- The decimal point in the values for Score occupies a column.

Numeric variables can be 40 columns wide, and string (text) variables can be up to 255 columns wide. A line can contain up to 1,024 columns. Because the specific columns containing data are defined the same way for each case, you can choose not to leave spaces between the variables; however, this practice is not normally recommended. Eliminating spaces increases the available room for data on the line, but readability suffers dramatically.

Make sure you start each line of data in column 1 (against the left margin). There should be no blank lines at the beginning or end of the datafile. You must save the output from the text editor or word processor in text (ASCII) form. If you save the datafile with the extension .txt or .dat, SPSS will automatically recognize it as a text datafile.

DELIMITED FORMAT

In the delimited format, values for each variable are delimited (separated) by a text or control character. Characters automatically recognized by the Text Import Wizard are tab, space, comma, and semicolon. SPSS allows the use of one or a

FIGURE A.2 Four-Case Dataset Delimited with Commas

combination of these characters as "delimiters." You can use and specify other characters as well.

Figure A.2 shows the dataset from the first example delimited with commas. You could also use the tab and semicolon characters as delimiters with this dataset. The Space character does not work as a delimiter here because the Text Import Wizard identifies the spaces between the text elements of the variable Name as delimiters and divides the name into separate values. In the second case for the variable Age, consecutive delimiters (,,) are used to define missing data.

LOADING TEXT DATAFILES WITH THE Text Import Wizard

From the menu bar, choose File>Read Text Data. Enter the filename of your text datafile, and click on the Open button to start the Text Import Wizard. "Wizards" will guide you through the various procedures, so a detailed description of the import process is not provided here. However, during the course of the procedure, you will be asked to specify the following:

- The filename
- The data format (fixed-width or delimited)
- Whether you entered variable names
- What line the first case begins on
- How many lines represent a case (or how many variables represent a case)
- How many cases to import
- Whether to save the file format (specifications) to be used again
- Whether to paste (use) SPSS syntax for the imported operation

You will also have the option of saving the imported specifications for later use with similar datafiles. If there is some ambiguity as to where one variable ends and the next begins, you may be asked to manually separate (parse) the variables.

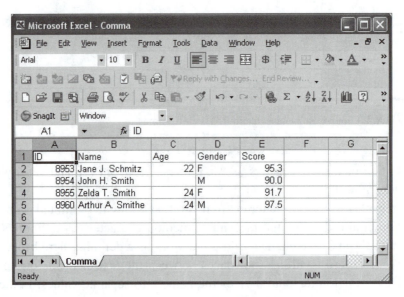

FIGURE A.3   Four-Case Dataset Created in an Excel Spreadsheet

If you have a preexisting text datafile created in Freefield format, you may have to organize it as a delimited datafile to import it with the Text Import Wizard. Optionally, you can create and run an SPSS syntax file to import a Freefield text datafile. If you have any problems, consult the Help menu.

USING A SPREADSHEET TO ENTER DATA

SPSS will read data directly from Microsoft Excel and Lotus files. Spreadsheet data entry is similar to fixed-width format. Variables are arranged in columns, and cases are represented by rows. Variable names can be entered on the first row of the spreadsheet above the columns that contain the values related to the variable. When the dataset used in the previous examples is created in an Excel spreadsheet, it appears as shown in Figure A.3.

Loading Data from a Spreadsheet into SPSS

To load data from a spreadsheet into SPSS, do the following:

1. From the menu bar, choose File>Open>Data.
2. Enter the filename.
3. Choose Excel (*.xls) or Lotus (*.w*) under the Files of type: drop list.
4. Click on the Open button.

5. Click on the Read variable names from the first row of data box if you have entered the variable names in the first row.

6. Enter the range of spreadsheet cells to load.

7. Click on the OK button.

USING A DATABASE TO ENTER DATA

SPSS will read data from any database program that produces dBASE (.dbf) files. Field names are read as variable names. If you are creating the database solely for the purpose of data entry, you should use variable names no longer than eight characters. If the field name is longer than eight characters, SPSS truncates it to eight. And if truncation produces duplicate variable names, duplicate variables are dropped!

Loading Data from a Database into SPSS

To load data from a database into SPSS, do the following:

1. Choose File>Open>Data from the menu bar.

2. Enter the filename.

3. Choose dBASE (.dbf) under the Files of type: drop list.

4. Click on the Open button.

If you have any problems, consult the Help menu.

USING THE Database Wizard

The Database Wizard allows retrieval of data from programs that use the Open DataBase Connectivity (ODBC) protocol. ODBC-compliant data sources include Microsoft Access and Excel. The Database Wizard may be accessed from the SPSS opening dialog box described in Assignment 1 (see Figure 1.4) by selecting the Create new query using the Database Wizard option. It can also be started from the SPSS File menu by choosing the Open Database option. The Database Wizard operates in a question-answer format.

Linking and Loading Data from a Microsoft Access File into SPSS

To retrieve data from a Microsoft Access database file using the Database Wizard, do the following:

1. From the menu bar, choose File>Open Database>New Query.

2. At the initial Database Wizard screen, click on MS Access Database and then on the Add Data Source button.

3. In the User Data Sources box, double-click on MS Access Database.

4. In the ODBC Microsoft Access Setup window, click on the Select button.

5. In the Select Database window, select the directory that contains your Access file in the Directories box to the right, and then click on the appropriate filename in the Database Name box on the left.

6. Navigate back to the Database Wizard screen by clicking on OK>OK>OK.

7. At the Database Wizard screen, click on MS Access Database and then on the Next> button.

8. If you want to move all the fields in the database into SPSS, click on and drag the entire table from the box on the left to the box on the right. To select a subset of fields from the data table, expand the table in the left-hand box by clicking on the "+" to the left of the table name. Click and drag the desired fields from the left-hand box to the right-hand box. When all the fields have been selected, click on the Next> button. From this point on you have the option of clicking on the Finish button if you do not need to further process the data in the fields you have selected.

9. You may now limit or filter the values of each of the fields you selected in Step 4. After this is completed, continue by clicking on the Next. button. You can add or change the variable names in the fifth screen. Click on Next> to continue.

10. The last screen displays the data query in SPSS syntax. You must run this query to load your data. Click on Finish to run the query and load the selected data. You may also save the query in an SPSS *.spq file by typing a filename in the Save query to file box. Using a saved query speeds up the query process. The next time you want to load these same data, choose File>Open Database>Run Query from the menu bar.

DATASETS

The datasets ALZ (Assignment 5), CardioStudy (Assignment 6), BABE (Assignment 7), and EduWage (Assignments 8 and 9), are shown on pages 88–89. You may use them to practice data entry using the SPSS Data Editor or the techniques described in this appendix to load data from text, spreadsheet, or database files.

ALZ Dataset			CardioStudy Dataset					
ID	PRE-MEM	POST-MEM	ID	AGEGR	CHOLCNT	ID (cont.)	AGEGR (cont.)	CHOLCNT (cont.)
1	21	23	1	2	249	41	1	180
2	18	11	2	3	245	42	2	219
3	29	29	3	2	238	43	2	222
4	42	34	4	2	243	44	1	201
5	21	14	5	3	238	45	1	168
6	37	38	6	2	238	46	2	214
7	23	24	7	3	263	47	2	234
8	15	21	8	3	256	48	3	253
9	29	24	9	3	256	49	1	185
10	35	41	10	1	173	50	2	248
11	18	21	11	2	254	51	2	167
12	37	33	12	3	261	52	1	161
13	23	19	13	1	209	53	1	171
14	18	20	14	3	262	54	3	266
15	23	31	15	3	251	55	2	254
16	42	42	16	3	264	56	3	173
17	21	22	17	1	130	57	1	187
18	28	40	18	3	264	58	1	182
19	15	26	19	3	239	59	3	259
20	18	12	20	2	259	60	1	188
21	29	32	21	3	205			
22	42	37	22	2	255			
23	21	17	23	2	203			
24	37	41	24	3	254			
25	23	27	25	2	234			
26	14	24	26	1	168			
27	29	27	27	3	248			
28	40	44	28	1	151			
29	17	24	29	1	151			
30	37	36	30	1	203			
31	23	22	31	3	213			
32	18	25	32	2	167			
33	24	34	33	1	173			
34	36	45	34	3	182			
35	35	43	35	2	249			
36	42	40	36	2	239			
37	21	20	37	1	190			
38	37	42	38	1	201			
39	20	29	39	2	241			
40	21	25	40	1	198			

BABE Dataset				EduWage Dataset			
ID	OVAL	RAND	FACE	ID	EDU	WAGE	WEX
1	180	170	340	1	15	12420	1.17
2	50	70	60	2	12	8880	27.00
3	70	120	170	3	12	8640	.00
4	40	60	90	4	17	27500	3.17
5	80	90	60	5	18	31300	3.92
6	80	60	180	6	12	10980	14.42
7	60	80	140	7	8	10080	28.67
8	50	70	120	8	8	7860	18.50
9	110	120	230	9	8	8940	14.00
10	60	70	120	10	12	12300	4.67
11	50	100	90	11	15	12660	5.33
12	100	90	150	12	12	8400	6.83
13	160	140	150	13	12	9360	.00
14	140	120	90	14	8	12780	34.00
15	180	170	340	15	16	18100	3.00
16	50	70	60	16	15	10980	4.00
17	70	120	170	17	19	36800	10.58
18	40	60	90	18	12	12300	25.58
				19	12	9780	8.92
				20	15	8280	10.08
				21	15	14820	.42
				22	15	13560	8.00
				23	15	12660	.83
				24	15	12840	5.58
				25	15	15660	3.83
				26	16	14400	3.83
				27	12	10740	4.00
				28	17	32500	1.50
				29	12	14280	6.00
				30	12	9240	1.08